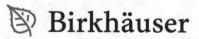

Frontiers in Mathematics

This series is designed to be a repository for up-to-date research results which have been prepared for a wider audience. Graduates and postgraduates as well as scientists will benefit from the latest developments at the research frontiers in mathematics and at the "frontiers" between mathematics and other fields like computer science, physics, biology, economics, finance, etc. All volumes are online available at SpringerLink.

More information about this series at http://www.springer.com/series/5388

Matej Brešar

Zero Product Determined Algebras

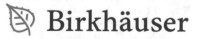 Birkhäuser

Matej Brešar
Faculty of Mathematics and Physics
University of Ljubljana
Ljubljana, Slovenia

Faculty of Natural Sciences and Mathematics
University of Maribor
Maribor, Slovenia

ISSN 1660-8046   ISSN 1660-8054   (electronic)
Frontiers in Mathematics
ISBN 978-3-030-80241-7   ISBN 978-3-030-80242-4   (eBook)
https://doi.org/10.1007/978-3-030-80242-4

Mathematics Subject Classification: 15A86, 16N40, 16P10, 16R60, 16S50, 16U40, 16W10, 16W20, 16W25, 17A01, 43A20, 46H05, 46H70, 46J10, 46L05, 47B47, 47B48, 47B49

This book is published under the imprint Birkhäuser, www.birkhauser-science.com, by the registered company Springer Nature Switzerland AG.
The registered company address is: Gewerbestrasse 11, 6330 Cham, Switzerland

Preface

An algebra A is said to be zero product determined (zpd for short) if every bilinear functional φ on A with the property that $\varphi(x, y) = 0$ whenever $xy = 0$ is of the form $\varphi(x, y) = \tau(xy)$ for some linear functional τ. If A is a Banach algebra, then we also require that φ and τ are continuous.

Perhaps this definition does not tell much. The intuition behind it is that many properties of A are determined by pairs of elements whose product is zero. This may still sound vague. The point, however, is that zpd (Banach) algebras form a rather wide class of not necessarily associative (Banach) algebras in which problems of different kinds can be solved. Their theory has reached a certain level of maturity. The purpose of this short book is to present it in a unified and concise manner, accessible to researchers and students of different backgrounds. The material is mostly taken from numerous papers published over the last 15 years, but some new results and new proofs of known results are also included.

The concept of a zpd algebra arose from two unrelated papers, [53] and [4]. The first one, written jointly with P. Šemrl, studies commutativity preserving linear maps on central simple algebras, and the second one, written jointly with J. Alaminos, J. Extremera, and A. Villena, studies local derivations and related maps on C^*-algebras. Their common feature is that the results on certain linear maps were derived from the consideration of bilinear maps that vanish on pairs of elements whose product is zero. This eventually led to the introduction of the concept of a zpd algebra and its systematic study, which evolved in two parallel directions: the algebraic and the (in many ways richer) analytic. Some terminological misfortunes occurred on the way: the analytic branch mostly uses the term "Banach algebra with property $\mathbb{B}$" (introduced in [6], in conjunction with the related property $\mathbb{A}$) for the concept that is similar to that of a "zero product determined algebra" used in the algebraic branch. One of the aims of the book is to unify both branches, and along the way we will make some terminological adjustments. Property $\mathbb{B}$ will therefore play only an auxiliary role.

The book is divided into three parts. Part I considers the algebraic theory and Part II the analytic theory. Part III is devoted to applications, that is, it examines problems from different areas of mathematics for which the zpd concept has proved useful. Some results in this last part are stated without detailed proofs. This is because the topics treated are very diverse and so many of them would need their own technical introduction. In the course of writing I realized that it is better to refer to original sources for details and focus primarily on ideas that illustrate the usefulness of the concepts studied in this book.

Parts of the book present results obtained jointly with my colleagues Jeronimo Alaminos, Jose Extremera, and Armando Villena from the University of Granada. I would like to thank them for the fruitful and pleasant long-term collaboration (and for being wonderful hosts during my frequent, enjoyable visits to Granada). The idea for the book arose from discussions with Armando, who was the leading force in developing the analytic branch of the theory. Although he did not join me as a coauthor, his influence can be felt as he has been giving me continuous help in the process of writing. My special thanks to him!

Finally, I would also like to thank Žan Bajuk for pointing out some errors in an earlier version, and the referees for useful comments.

Ljubljana and Maribor, Slovenia   Matej Brešar
April 2021

Contents

Part I

Algebraic Theory

Zero Product Determined Nonassociative Algebras

Part I is devoted to a purely algebraic consideration of the class of zero product determined algebras. We will be mainly concerned with the question of which algebras do, and which do not, belong to this class.

This first chapter is of a preliminary nature. We will introduce the basic concepts and discuss their elementary properties.

1.1   The Definition of a zpd Nonassociative Algebra

Throughout the chapter, A stands for a *nonassociative* algebra over a field F. This does not mean that the product in A is not associative, but it is *not necessarily associative*. Associative algebras are thus special (and in fact the most important) examples of nonassociative algebras. We are also not assuming that A is commutative and neither that A has a unity 1. Whenever the latter will be assumed, we will speak about a *unital algebra*.

Let τ be a linear functional on A. Define $\varphi : A \times A \to F$ by

$$\varphi(x, y) = \tau(xy). \tag{1.1}$$

Obviously, φ is a *bilinear functional on A*, that is, it is linear in each variable separately. Bilinear functionals of the form (1.1) appear naturally in mathematics. Incidentally, some readers may be familiar with the notion of a Frobenius algebra, which is defined as a finite-dimensional associative unital algebra endowed with a non-degenerate bilinear functional φ of the form (1.1) (for some linear functional τ). However, we will not be interested in non-degeneracy and related properties of bilinear functionals. What is important for us is

© The Author(s), under exclusive license to Springer Nature Switzerland AG 2021
M. Brešar, *Zero Product Determined Algebras*, Frontiers in Mathematics,
https://doi.org/10.1007/978-3-030-80242-4_1

that every bilinear functional φ of the form (1.1) has the property that for all $x, y \in A$,

$$xy = 0 \implies \varphi(x, y) = 0. \tag{1.2}$$

In some algebras A, the bilinear functionals of the form (1.1) are also the only bilinear functionals satisfying (1.2). These are the algebras from the title of the book.

Definition 1.1 A nonassociative algebra A over a field F is said to be *zero product determined* if, for every bilinear functional $\varphi : A \times A \to F$ satisfying $\varphi(x, y) = 0$ whenever $x, y \in A$ are such that $xy = 0$, there exists a linear functional τ on A such that $\varphi(x, y) = \tau(xy)$ for all $x, y \in A$.

Throughout the book, we will use *zpd* as an abbreviation for "zero product determined." We will thus usually say that an algebra A satisfying the condition of Definition 1.1 is zpd.

The term "zero product determined algebra" was used for the first time in the 2009 paper [58]. The concept as such, however, implicitly appeared in some earlier works. Of course, it is not immediately clear why this concept should be of interest; showing that it is what this book is about.

Remark 1.2 Throughout the book, we write

$$A^2 = \operatorname{span}\{xy \mid x, y \in A\},$$

i.e., A^2 is the linear span of all elements that can be written as a product of two elements in A. In Definition 1.1, it would be enough to require that τ is defined only on A^2 (rather than on A). However, we can always extend a linear functional from a subspace to the whole space, so these two conditions are equivalent.

The next proposition provides some equivalent definitions.

Proposition 1.3 *Let A be a nonassociative algebra over a field F. The following conditions are equivalent:*

  (i) *A is zpd.*
 (ii) *If a bilinear functional $\varphi : A \times A \to F$ has the property that $\varphi(x, y) = 0$ whenever $x, y \in A$ are such that $xy = 0$, then $\sum_i \varphi(x_i, y_i) = 0$ whenever $x_i, y_i \in A$ are such that $\sum_i x_i y_i = 0$.*
(iii) *If $z_i, w_i \in A$ are such that $\sum_i z_i w_i = 0$, then the element $\sum_i z_i \otimes w_i$ from $A \otimes A$ lies in $\operatorname{span}\{x \otimes y \in A \otimes A \mid xy = 0\}$.*
(iv) *For every vector space X over F and every bilinear map $\varphi : A \times A \to X$ satisfying $\varphi(x, y) = 0$ whenever $x, y \in A$ are such that $xy = 0$, there exists a linear map $T : A \to X$ such that $\varphi(x, y) = T(xy)$ for all $x, y \in A$.*

Proof (i) $\Longleftrightarrow$ (ii). It is obvious that (i) implies (ii). To prove the converse, take a bilinear functional φ satisfying $\varphi(x, y) = 0$ whenever $xy = 0$. Assuming that (ii) holds, it follows that $\tau : A^2 \to F$ given by

$$\tau\left(\sum_i x_i y_i\right) = \sum_i \varphi(x_i, y_i)$$

is a well-defined linear functional. Obviously, τ satisfies $\varphi(x, y) = \tau(xy)$, and so, in light of Remark 1.2, A is zpd.

(ii) $\Longleftrightarrow$ (iii). Denote by U the linear span of all elements $x \otimes y$ in $A \otimes A$ such that $xy = 0$, and by V the linear space of all $\sum_i z_i \otimes w_i$ in $A \otimes A$ such that $\sum_i z_i w_i = 0$. Trivially, $U \subseteq V$. We have to show that $U = V$ if and only if A satisfies (ii).

Take a bilinear functional φ on A satisfying $\varphi(x, y) = 0$ whenever $xy = 0$. Let Ψ be the linear functional on $A \otimes A$ given by

$$\Psi(x \otimes y) = \varphi(x, y).$$

Obviously, Ψ vanishes on U. If $U = V$, then it follows from $\sum_i z_i w_i = 0$ that

$$\sum_i \varphi(z_i, w_i) = \Psi\left(\sum_i z_i \otimes w_i\right) = 0.$$

Thus, $U = V$ implies (ii).

Suppose $U \subsetneq V$. Take a linear functional Ψ on $A \otimes A$ such that $\Psi(U) = \{0\}$ and $\Psi(V) \neq \{0\}$. Then $\varphi : A \times A \to F$ defined by

$$\varphi(x, y) = \Psi(x \otimes y)$$

satisfies $\varphi(x, y) = 0$ whenever $xy = 0$, but there exist $z_i, w_i \in A$ such that $\sum_i z_i w_i = 0$ and

$$0 \neq \Psi\left(\sum_i z_i \otimes w_i\right) = \sum_i \varphi(z_i, w_i).$$

This means that (ii) does not hold.

(ii) $\Longleftrightarrow$ (iv). We only have to prove that (ii) implies (iv), since the converse is trivial. Assume, therefore, that (ii) holds and that a bilinear map $\varphi : A \times A \to X$ satisfies $\varphi(x, y) = 0$ whenever $xy = 0$. Take a linear functional γ on X and note that the condition (ii) is applicable to the composition $\gamma \circ \varphi : A \times A \to F$. Therefore, if $x_i, y_i \in A$ are such that $\sum_i x_i y_i = 0$, then

$$\gamma\left(\sum_i \varphi(x_i, y_i)\right) = \sum_i (\gamma \circ \varphi)(x_i, y_i) = 0.$$

Since γ is an arbitrary linear functional on A, it follows that $\sum_i \varphi(x_i, y_i) = 0$. The rest of the proof is the same as that of (ii) $\Longrightarrow$ (i). Indeed, we first note that $T : A^2 \to X$ given by

$$T\left(\sum_i x_i y_i\right) = \sum_i \varphi(x_i, y_i)$$

is a well-defined linear map satisfying $\varphi(x, y) = T(xy)$, and then just extend T to a linear map from A to X. $\qquad\square$

Conditions (ii) and (iii) provide two ways for proving that an algebra A is zpd. Condition (iv), on the other hand, is often of essential importance when one is in a position to apply the condition that A is zpd. Here is an illustration.

Proposition 1.4 *Let A be a unital nonassociative algebra. Suppose that the vector space of A is an algebra under another multiplication, $*$, such that $x * y = 0$ whenever $xy = 0$. If A is zpd, then $x * y = 1 * xy$ and $1 * x = x * 1$ for all $x, y \in A$.*

Proof The preceding proposition, more precisely the implication (i) $\Longrightarrow$ (iv), shows that there is a linear map $T : A \to A$ such that $x * y = T(xy)$ for all $x, y \in A$. Setting $x = 1$ we obtain $1 * y = T(y)$, and hence $x * y = T(xy) = 1 * xy$. Similarly, setting $y = 1$ gives $x * 1 = T(x)$, and so $1 * x = x * 1$. $\qquad\square$

This simple result is supposed to serve just as a small justification for choosing the name "zero product determined." Later, we will see that indeed many properties of zpd algebras are determined by pairs of elements whose product is zero.

We assumed that A is unital primarily to make the result more transparent. It is often the case that the presence of unity makes the study of zpd algebras much easier. The next corollary shows that conditions (ii) and (iii) from Proposition 1.3 can be considerably simplified if A is unital.

Corollary 1.5 *Let A be a unital nonassociative algebra over a field F. The following conditions are equivalent:*

(i) *A is zpd.*
(ii) *If a bilinear functional $\varphi : A \times A \to F$ has the property that $\varphi(x, y) = 0$ whenever $x, y \in A$ are such that $xy = 0$, then*

$$\varphi(x, y) = \varphi(xy, 1) \quad (x, y \in A).$$

(iii) *For all $z, w \in A$, $z \otimes w - zw \otimes 1 \in \text{span}\,\{x \otimes y \in A \otimes A \mid xy = 0\}$.*

Proof It is clear that (1.1) implies $\varphi(x, 1) = \tau(x)$, and hence $\varphi(x, y) = \varphi(xy, 1)$. Thus, (i) and (ii) are equivalent. Since (iii) is a special case of condition (iii) of Proposition 1.3, it only remains to show that (iii) implies (ii). The proof of this is essentially a repetition of the proof of the implication (iii) $\Longrightarrow$ (ii) of Proposition 1.3. Take a bilinear functional φ on A such that $\varphi(x, y) = 0$ whenever $xy = 0$, and define a linear functional Ψ on $A \otimes A$ by

$$\Psi(x \otimes y) = \varphi(x, y).$$

Obviously, if x_i, y_i are such that $x_i y_i = 0$ for every i, then

$$\Psi\left(\sum_i x_i \otimes y_i\right) = 0.$$

Assuming that condition (iii) holds, it thus follows that for all $z, w \in A$,

$$\Psi(z \otimes w - zw \otimes 1) = 0,$$

i.e., $\varphi(z, w) = \varphi(zw, 1)$. $\square$

Besides the trivial example where $A = F$, it may not be obvious that zpd algebras actually exist. However, we leave this fundamental problem aside for now. It is easier to find non-examples. The next proposition shows that unital zpd algebras of nontrivial dimension must have an abundance of zero-divisors. But first we record a simple lemma on tensor products, which is nothing but a convenient restatement of the fundamental property that $\sum_i u_i \otimes v_i = 0$ with u_i linearly independent implies $v_i = 0$ for each i (see [47, Lemmas 4.9 and 4.11]).

Lemma 1.6 *Let U and V be vector spaces over a field F. Suppose that $u_i, z_j \in U$ and $v_i, w_j \in V$ satisfy*

$$\sum_i u_i \otimes v_i = \sum_j z_j \otimes w_j.$$

If the vectors u_i are linearly independent, then each v_i lies in the linear span of the vectors w_j. Similarly, if the vectors v_i are linearly independent, then each u_i lies in the linear span of the vectors z_j.

By a *left zero-divisor* in an algebra A we mean a nonzero element $x \in A$ such that $xy = 0$ for some nonzero $y \in A$. A *right zero-divisor* is defined analogously. A *zero-divisor* is an element that is either a left or right zero-divisor.

Proposition 1.7 *If A is a unital zpd nonassociative algebra of dimension greater than 1, then every element in A is a sum of right zero-divisors (as well as a sum of left zero-divisors).*

Proof Take any $z \in A$ that is not a scalar multiple of the unity 1 of A. By condition (iii) of Corollary 1.5, there exist $x_i, y_i \in A$ such that $x_i y_i = 0$ and

$$z \otimes 1 - 1 \otimes z = \sum_i x_i \otimes y_i.$$

Lemma 1.6 shows that both z and 1 lie in the linear span of the elements y_i (as well as in the linear span of the elements x_i). Since a scalar multiple of a right (resp. left) zero-divisor is either zero or a right (resp. left) zero-divisor, this yields the desired conclusion.   □

Corollary 1.8 *Let A be a unital nonassociative algebra of dimension greater than 1. If A has no zero-divisors, then A is not zpd.*

The true motivation for introducing and studying zpd algebras lies in their applications; more precisely, they provide a convenient setting in which various problems from different mathematical areas can be effectively solved. This will be presented in Part III. However, the reader might appreciate seeing some evidence of the usefulness of the zpd concept already now. The following three briefly presented examples are hopefully illustrative. They will be discussed in greater detail in Part III (more specifically, in Sects. 7.1, 7.2, 8.2, and 9.1).

Example 1.9 A linear map T from an algebra A to an algebra B is said to be *zero product preserving* if $T(x)T(y) = 0$ whenever $xy = 0$. Describing the form of such a map is a well-studied problem. One usually wishes to show that T is close to a homomorphism. The assumption that A is zpd immediately brings us close to this goal. Indeed, by Proposition 1.3 (iv) there exists a linear map $S : A \to B$ such that $T(x)T(y) = S(xy)$ for all $x, y \in A$. If we also assume that A and B are unital and $T(1) = 1$, then, by taking $y = 1$, we get $S = T$ and so T is a homomorphism. Without this additional assumption, the problem is much more intriguing. It is especially interesting in the case where A and B are Lie algebras (the assumption that T preserves zero products can then be read as that T preserves commutativity).

Example 1.10 A linear map $d : A \to A$ is called a *derivation* if $d(xy) = d(x)y + xd(y)$ for all $x, y \in A$. A *local derivation* is a linear map $\delta : A \to A$ such that, for every $x \in A$, there exists a derivation d_x of A such that $\delta(x) = d_x(x)$. The standard question is whether every local derivation of A is a derivation. Let us show that the answer is positive if A is a unital zpd associative algebra. We start the proof by noticing that if $x, y, z \in A$ satisfy $xy = yz = 0$, then

$$x\delta(y)z = xd_y(y)z = d_y(xy)z - d_y(x)yz = 0.$$

Now fix $y', z' \in A$ such that $y'z' = 0$ and define $\varphi_1 : A \times A \to A$ by

$$\varphi_1(x, y) = x\delta(yy')z'.$$

If $x, y \in A$ satisfy $xy = 0$, then $x(yy') = (yy')z' = 0$ and hence, as observed above, $\varphi_1(x, y) = 0$. From Proposition 1.3 (iv) we thus see that φ_1 in particular satisfies $\varphi_1(1, x) - \varphi_1(x, 1) = 0$ for all $x \in A$. That is,

$$\delta(xy')z' - x\delta(y')z' = 0$$

for any x and any pair y', z' satisfying $y'z' = 0$. Hence, the map $\varphi_2 : A \times A \to A$ defined by

$$\varphi_2(y', z') = \delta(xy')z' - x\delta(y')z'$$

(with x arbitrary but fixed) satisfies $\varphi_2(y, 1) = \varphi_2(1, y)$ for all $y \in A$, i.e.,

$$\delta(xy) - x\delta(y) = \delta(x)y - x\delta(1)y$$

for all $y \in A$, as well as all $x \in A$. From

$$\delta(1) = d_1(1) = d_1(1^2) = d_1(1)1 + 1d_1(1) = 2\delta(1)$$

we see that $\delta(1) = 0$, and so δ is a derivation.

Example 1.11 Let A be an associative algebra. Is every *commutator*, i.e., an element of the form $xy - yx$ for some $x, y \in A$, a sum of elements whose square is zero? This question does not always have a positive answer, but, as shown by various authors through the decades, for many algebras A it does. Let us show that any unital zpd algebra A has this property. Denote by N the set of all elements that can be written as sums of square-zero elements. Obviously, N is a vector subspace of A. Consider the bilinear map $\varphi : A \times A \to A/N$ given by $\varphi(x, y) = yx + N$. Note that $xy = 0$ implies $(yx)^2 = 0$ and hence $\varphi(x, y) = 0$. Using Proposition 1.3 (iv) it follows, just as in the proof of Corollary 1.5, that φ satisfies $\varphi(x, y) = \varphi(xy, 1)$ for all $x, y \in A$. This means that $xy - yx$ lies in N, as desired.

1.2   Symmetrically zpd Nonassociative Algebras

A bilinear functional φ is said to be *symmetric* if

$$\varphi(x, y) = \varphi(y, x) \quad (x, y \in A).$$

If A is commutative and φ is of the form $\varphi(x, y) = \tau(xy)$ for some linear functional τ on A, then φ is symmetric. The following modification of Definition 1.1 is therefore natural.

Definition 1.12 A commutative nonassociative algebra A over a field F is said to be *symmetrically zero product determined*, or *symmetrically zpd* for short, if, for every symmetric bilinear functional $\varphi : A \times A \to F$ satisfying $\varphi(x, y) = 0$ whenever $x, y \in A$ are such that $xy = 0$, there exists a linear functional τ on A such that $\varphi(x, y) = \tau(xy)$ for all $x, y \in A$.

Of course, if a commutative algebra is zpd, then it is also symmetrically zpd. Example 2.23 below shows that the converse does not hold.

An analogue of Corollary 1.5 for symmetrically zpd (commutative) algebras reads as follows.

Proposition 1.13 *Let A be a commutative unital nonassociative algebra over a field F of characteristic not 2. The following conditions are equivalent:*

(i) *A is symmetrically zpd.*
(ii) *If a symmetric bilinear functional $\varphi : A \times A \to F$ has the property that $\varphi(x, y) = 0$ whenever $x, y \in A$ are such that $xy = 0$, then*

$$\varphi(x, x) = \varphi(x^2, 1) \quad (x \in A).$$

(iii) *For each $z \in A$, there exist x_i, $y_i \in A$ such that*

$$2z \otimes z - z^2 \otimes 1 - 1 \otimes z^2 = \sum_i x_i \otimes y_i$$

and $x_i y_i = 0$ for every i.

Proof If a symmetric bilinear functional φ on A satisfies $\varphi(x, x) = \varphi(x^2, 1)$, then it also satisfies $\varphi(x, y) = \varphi(xy, 1)$. Indeed, since A is commutative and the characteristic of F is not 2, this follows immediately by replacing x by $x + y$ in $\varphi(x, x) = \varphi(x^2, 1)$. Therefore, just as in the proof of Corollary 1.5, we see that (i) and (ii) are equivalent.

As in the proof of Proposition 1.3, we set

$$U = \text{span}\,\{x \otimes y \in A \otimes A \mid xy = 0\}.$$

Assuming that (iii) does not hold, it follows that there exists a linear functional Ψ on $A \otimes A$ such that $\Psi(U) = \{0\}$ and

$$\Psi\left(2z \otimes z - z^2 \otimes 1 - 1 \otimes z^2\right) \neq 0$$

for some $z \in A$. Define $\varphi : A \times A \to F$ by

$$\varphi(x,\,y) = \Psi(x \otimes y + y \otimes x).$$

It is clear that φ is a symmetric bilinear functional that vanishes on every pair of elements whose product is 0. However, since $\varphi(z, z) \neq \varphi(z^2, 1)$, φ does not satisfy (ii).

Finally, assume that (iii) holds and that φ is a symmetric bilinear functional on A that vanishes on pairs of elements whose product is 0. The linear functional Ψ on $A \otimes A$ defined by

$$\Psi(x \otimes y) = \varphi(x,\,y)$$

then vanishes on U. Since, by assumption, $2z \otimes z - z^2 \otimes 1 - 1 \otimes z^2$ lies in U for every $z \in A$, it follows that

$$\Psi\left(2z \otimes z - z^2 \otimes 1 - 1 \otimes z^2\right) = 0 \quad (z \in A).$$

That is,

$$2\varphi(z, z) = \varphi(z^2, 1) + \varphi(1, z^2) \quad (z \in A).$$

As φ is symmetric, this proves (ii). $\qquad\qquad\qquad\qquad\qquad\qquad\qquad\qquad\quad \square$

1.3 Stability Under Algebraic Constructions

In this section, we will show that the class of zpd algebras is stable under some algebraic constructions. These results may, at present, seem vacuous since we have not yet provided nontrivial examples of zpd algebras. However, the problem of finding examples is subtle, while the problems that will be addressed now are of a formal nature, enabling us to continue working in the context of general nonassociative algebras. This simply means that, like in the preceding sections, we will avoid multiplying more than two elements together.

We start with homomorphic images.

Theorem 1.14 *Let A be a zpd nonassociative algebra and let $\theta : A \to B$ be an algebra epimorphism. If $A^2 \cap \ker \theta = A \ker \theta$, then B is a zpd algebra too.*

Proof Let $\varphi : B \times B \to F$ be a bilinear functional satisfying $\varphi(u, v) = 0$ whenever $uv = 0$. Define $\overline{\varphi} : A \times A \to F$ by

$$\overline{\varphi}(x, y) = \varphi(\theta(x), \theta(y)).$$

Obviously, $xy = 0$ implies $\theta(x)\theta(y) = 0$ and hence $\overline{\varphi}(x, y) = 0$. Since A is zpd, there exists a linear functional $\overline{\tau}$ on A such that

$$\overline{\varphi}(x, y) = \overline{\tau}(xy)$$

for all $x, y \in A$. Now let $\tau_0 : B^2 \to F$ be given by

$$\tau_0\left(\sum_i \theta(x_i)\theta(y_i)\right) = \overline{\tau}\left(\sum_i x_i y_i\right).$$

To show that τ_0 is well-defined, assume that $\sum_i \theta(x_i)\theta(y_i) = 0$ for some $x_i, y_i \in A$. Then $\sum_i x_i y_i \in A^2 \cap \ker \theta$. Therefore, by the assumption of the theorem, there exist $z_j \in A$ and $k_j \in \ker \theta$ such that $\sum_i x_i y_i = \sum_j z_j k_j$. Hence,

$$\overline{\tau}\left(\sum_i x_i y_i\right) = \overline{\tau}\left(\sum_j z_j k_j\right)$$

$$= \sum_j \overline{\varphi}(z_j, k_j)$$

$$= \sum_j \varphi(\theta(z_j), \theta(k_j))$$

$$= 0$$

since $\theta(k_j) = 0$. This proves that τ_0 is well-defined. If τ is any linear functional on B extending τ_0, then

$$\varphi(\theta(x), \theta(y)) = \overline{\varphi}(x, y) = \overline{\tau}(xy) = \tau(\theta(x)\theta(y)),$$

which proves that B is zpd. □

The condition $A^2 \cap \ker \theta = A \ker \theta$ is trivially fulfilled if A is a unital algebra. The following corollary therefore holds.

Corollary 1.15 *If A is a unital zpd nonassociative algebra, then so is any homomorphic image of A.*

The *direct product* of the family of algebras $(A_i)_{i \in I}$ is the Cartesian product $\Pi_{i \in I} A_i$ endowed with the componentwise operations. Its subalgebra consisting of all $(x_i) \in \Pi_{i \in I} A_i$ such that $x_i = 0$ for all but a finite number of $i \in I$ is called the *direct sum* of the family $(A_i)_{i \in I}$ and is denoted by $\oplus_{i \in I} A_i$. Obviously, $\oplus_{i \in I} A_i = \Pi_{i \in I} A_i$ if and only if I is a finite set. If $I = \{1, \ldots, n\}$, then we write this algebra as $A_1 \times \cdots \times A_n$ or $A_1 \oplus \cdots \oplus A_n$.

The next theorem concerns direct sums. Example 2.24 in the next chapter shows that the direct product case is different.

Theorem 1.16 *Let $(A_i)_{i \in I}$ be a family of nonassociative algebras. Then their direct sum $\oplus_{i \in I} A_i$ is zpd if and only if A_i is zpd for every $i \in I$.*

Proof Suppose $\oplus_{i \in I} A_i$ is zpd. Pick $j \in I$ and let $\theta : \oplus_{i \in I} A_i \to A_j$ be the canonical epimorphism (defined by $\theta((x_i)) = x_j$). Observe that θ satisfies the assumption of Theorem 1.14, and so A_j is zpd.

Assume now that each A_i, $i \in I$, is zpd, and let us prove that $A = \oplus_{i \in I} A_i$ is zpd. Let φ be a bilinear functional on A satisfying $\varphi(x, y) = 0$ whenever $xy = 0$. For each $a_i \in A_i$, we write $\widetilde{a}_i$ for the element $(x_\ell) \in A$ defined by $x_i = a_i$ and $x_\ell = 0$ for $\ell \neq i$. Clearly,

$$(a_i, b_i) \mapsto \varphi(\widetilde{a}_i, \widetilde{b}_i)$$

is a bilinear functional on A_i satisfying $\varphi(\widetilde{a}_i, \widetilde{b}_i) = 0$ whenever $a_i b_i = 0$. Since A_i is zpd, condition (ii) of Proposition 1.3 shows that for all $a_{i,j}, b_{i,j} \in A_i$,

$$\sum_j a_{i,j} b_{i,j} = 0 \implies \sum_j \varphi(\widetilde{a_{i,j}}, \widetilde{b_{i,j}}) = 0.$$

Now take $x_j, y_j \in A$ such that $\sum_j x_j y_j = 0$. Our goal is to prove that $\sum_j \varphi(x_j, y_j) = 0$. Write $x_j = (a_{i,j})$ and $y_j = (b_{i,j})$ with $a_{i,j}, b_{i,j} \in A_i$, and let $i_1, \ldots, i_m \in I$ be such that $a_{i,j} = b_{i,j} = 0$ for all j and all $i \notin \{i_1, \ldots, i_m\}$. The condition $\sum_j x_j y_j = 0$ can be equivalently written as

$$\sum_j a_{i_t,j} b_{i_t,j} = 0 \quad (t = 1, \ldots, m).$$

As observed above, this implies

$$\sum_j \varphi(\widetilde{a_{i_t,j}}, \widetilde{b_{i_t,j}}) = 0 \quad (t = 1, \ldots, m).$$

Note also that if $s \neq t$, then $\widetilde{a_{i_s,j}} \widetilde{b_{i_t,j}} = 0$ and hence $\varphi(\widetilde{a_{i_s,j}}, \widetilde{b_{i_t,j}}) = 0$. Consequently,

$$\sum_j \varphi(x_j, y_j) = \sum_j \varphi(\widetilde{a_{i_1,j}} + \cdots + \widetilde{a_{i_m,j}}, \widetilde{b_{i_1,j}} + \cdots + \widetilde{b_{i_m,j}})$$

$$= \sum_j \varphi(\widetilde{a_{i_1,j}}, \widetilde{b_{i_1,j}}) + \cdots + \varphi(\widetilde{a_{i_m,j}}, \widetilde{b_{i_m,j}})$$

$$= \sum_{t=1}^m \sum_j \varphi(\widetilde{a_{i_t,j}}, \widetilde{b_{i_t,j}}) = 0,$$

as desired.   □

Our next topic is tensor products. First we record an elementary lemma on tensor products of linear maps.

Lemma 1.17 *Let U_1, U_2, V_1, V_2 be vector spaces over a field F, and let $T_1 : U_1 \to V_1$ and $T_2 : U_2 \to V_2$ be linear maps. Then the kernel of $T_1 \otimes T_2 : U_1 \otimes U_2 \to V_1 \otimes V_2$ is equal to $\ker T_1 \otimes U_2 + U_1 \otimes \ker T_2$.*

Proof It is clear that $Z = \ker T_1 \otimes U_2 + U_1 \otimes \ker T_2$ is contained in $K = \ker(T_1 \otimes T_2)$. We must prove that every $k = \sum_{i=1}^n x_i \otimes y_i \in K$ lies in Z. We proceed by induction on n. The $n = 1$ case is trivial, so let $n > 1$. There is nothing to prove if each x_i lies in $\ker T_1$, so we may assume that $T_1(x_1) \neq 0$. Since $\sum_{i=1}^n T_1(x_i) \otimes T_2(y_i) = 0$, it follows from Lemma 1.6 that $T_2(y_1) = \sum_{j=2}^n \lambda_j T_2(y_j)$ for some $\lambda_j \in F$. We now write

$$k = x_1 \otimes \left(y_1 - \sum_{j=2}^n \lambda_j y_j \right) + \sum_{j=2}^n (x_j + \lambda_j x_1) \otimes y_j.$$

The first term on the right-hand side lies in $U_1 \otimes \ker T_2 \subseteq Z$, while the summation on the right-hand side lies in Z by the induction assumption. Thus, $k \in Z$.   □

Theorem 1.18 *If A_1 and A_2 are zpd nonassociative algebras, then so is $A_1 \otimes A_2$.*

Proof Let $\pi_i : A_i \otimes A_i \to A_i$ be linear maps given by $\pi_i(x \otimes y) = xy, i = 1, 2$. Note that condition (iii) of Proposition 1.3 can be described as that $\ker \pi_i$ is equal to

$$W_i = \text{span}\,\{x \otimes y \in A_i \otimes A_i \,|\, xy = 0\}, \quad i = 1, 2.$$

Set $A = A_1 \otimes A_2$. We must prove that the kernel of $\pi : A \otimes A \to A, \pi(x \otimes y) = xy$, is equal to

$$W = \text{span}\,\{x \otimes y \in A \otimes A \,|\, xy = 0\}.$$

Let

$$\sigma : A_1 \otimes A_2 \otimes A_1 \otimes A_2 \to A_1 \otimes A_1 \otimes A_2 \otimes A_2$$

be the canonical isomorphism defined by

$$\sigma(x_1 \otimes x_2 \otimes y_1 \otimes y_2) = x_1 \otimes y_1 \otimes x_2 \otimes y_2.$$

Then $\pi = (\pi_1 \otimes \pi_2) \circ \sigma$ and so $x \in \ker \pi$ if and only if $\sigma(x) \in \ker(\pi_1 \otimes \pi_2)$. By Lemma 1.17,

$$\ker(\pi_1 \otimes \pi_2) = W_1 \otimes (A_2 \otimes A_2) + (A_1 \otimes A_1) \otimes W_2.$$

Thus, if $x \in \ker \pi$, then $\sigma(x)$ is equal to a sum of elements of the form

$$x_1 \otimes y_1 \otimes x_2 \otimes y_2, \text{ with either } x_1 y_1 = 0 \text{ or } x_2 y_2 = 0,$$

and hence $x = \sigma^{-1}(\sigma(x))$ lies in W. Therefore, $\ker \pi \subseteq W$. The converse inclusion is trivial.   □

Corollary 2.17 below states that if A is any associative unital algebra and $n \geq 2$, then $M_n(A)$, the algebra of all $n \times n$ matrices with entries in A, is zpd, regardless of whether A is zpd or not. Since $M_n(A) \cong M_n(F) \otimes A$, this shows that the converse of Theorem 1.18 does not hold.

We conclude with a negative result, showing that the zpd property is not preserved by adjoining a unity to an algebra. The next proposition actually holds no matter whether A is zpd or not.

Proposition 1.19 *Let A be a nonassociative algebra such that $A^2 \neq A$. Then $A^{\#}$, the algebra obtained by adjoining a unity to A, is not zpd.*

Proof Consider the bilinear map φ from $A^\# \times A^\#$ to the quotient space $A^\#/A^2$ defined by

$$\varphi(\alpha + x, \beta + y) = \alpha y + A^2,$$

where α and β are scalars and x and y are elements in A. Suppose $(\alpha + x)(\beta + y) = 0$. Then $\alpha\beta = 0$ and $\alpha y + \beta x + xy = 0$, so either $\alpha = 0$ or $\beta = 0$ and in each case $\alpha y \in A^2$. Thus, φ vanishes on pairs of elements whose product is 0. However, if $z \in A \setminus A^2$, then $\varphi(1, z) = z + A^2 \neq 0$ while $\varphi(z, 1) = 0$, which shows that $A^\#$ does not satisfy condition (iv) of Proposition 1.3. Thus, $A^\#$ is not zpd.   □

A more detailed analysis of the topic of this section is given in [61].

Zero Product Determined Rings and Algebras

We now turn to associative algebras. The main message of this chapter is that, in these algebras, the zpd property is closely related to the property of being generated by idempotents.

We begin our consideration in the more general context of rings, then proceed to algebras, and at the end obtain a definitive result in the finite-dimensional case.

2.1 zpd Rings

Although our general framework is that of algebras over fields, we will now make a small exception and start this chapter by considering rings. By a ring we will mean an associative, but not necessarily commutative or unital ring.

In the next definition, we assume that the map φ has range in R. The main reason for this restrictive choice is to avoid tedious technicalities in the short discussion that follows.

Definition 2.1 A ring R is said to be *zero product determined*, or *zpd* for short, if, for every biadditive map $\varphi : R \times R \to R$ satisfying $\varphi(x, y) = 0$ whenever $x, y \in R$ are such that $xy = 0$, there exists an additive map $T : R \to R$ such that $\varphi(x, y) = T(xy)$ for all $x, y \in R$.

The next simple but crucial lemma has first appeared (in some form) in [46]; the idea of the proof, however, goes back to [67]. Recall that a ring element e is called an *idempotent* if $e^2 = e$.

© The Author(s), under exclusive license to Springer Nature Switzerland AG 2021
M. Brešar, *Zero Product Determined Algebras*, Frontiers in Mathematics,
https://doi.org/10.1007/978-3-030-80242-4_2

Lemma 2.2 *Let R be a ring and let $\varphi : R \times R \to R$ be a biadditive map such that $\varphi(x, y) = 0$ whenever $xy = 0$. Then the set*

$$R_0 = \{r \in R \mid \varphi(xr, z) = \varphi(x, rz) \text{ for all } x, z \in R\}$$

is a subring of R containing all idempotents in R.

Proof Showing that R_0 is a subring is straightforward. Let e be an idempotent in R. Then

$$xe \cdot (z - ez) = 0 \text{ and } (x - xe) \cdot ez = 0$$

for any $x, z \in R$, and so

$$\varphi(xe, z - ez) = 0 \text{ and } \varphi(x - xe, ez) = 0.$$

Writing these two identities as

$$\varphi(xe, z) = \varphi(xe, ez) \text{ and } \varphi(x, ez) = \varphi(xe, ez),$$

we see, by comparing, that $\varphi(xe, z) = \varphi(x, ez)$. That is, $e \in R_0$.   □

In particular, if the ring R is generated by its idempotents, then $R_0 = R$. This yields the following theorem.

Theorem 2.3 *If a unital ring R is generated by idempotents, then R is zpd.*

Proof Let φ be as in Lemma 2.2. Since $R_0 = R$, $\varphi(xy, z) = \varphi(x, yz)$ for all $x, y, z \in R$. Setting $z = 1$ we obtain $\varphi(xy, 1) = \varphi(x, y)$. This shows that R is zpd.   □

What are examples of unital rings generated by idempotents? In principle, we can take any unital ring and the unital subring generated by any set of idempotents in this ring provides such an example; if this set contains nontrivial idempotents (i.e., idempotents different from 0 and 1), then this example is nontrivial. However, this is purely theoretical. Let us point out two specific types of rings generated by idempotents.

As usual, $M_n(S)$ denotes the ring of all $n \times n$ matrices with entries in the ring S.

Corollary 2.4 *Let S be any unital ring and let $n \geq 2$. Then the ring $R = M_n(S)$ is generated by idempotents and is hence zpd.*

Proof We write e_{ij} for the (standard) *matrix units* in R. That is, e_{ij} is the matrix whose (i, j) entry is 1 and all other entries are 0. Further, by se_{ij} we denote the matrix whose (i, j) entry is $s \in S$ and all other entries are 0. Since every matrix in R is a sum of the matrices se_{ij}, it is enough to show that each such matrix can be expressed by idempotents. If $i \neq j$, then $e_{ii} + se_{ij}$ is an idempotent, so

$$se_{ij} = (e_{ii} + se_{ij}) - e_{ii}$$

is a difference of two idempotents. From $se_{ii} = se_{ij} \cdot e_{ji}$, where $j \neq i$, it thus follows that se_{ii} also lies in the subring generated by idempotents. $\square$

Let R_1 and R_2 be rings. Recall that M is said to be an (R_1, R_2)-*bimodule* if M is both a left R_1-module and right R_2-module, and

$$(r_1 \cdot m) \cdot r_2 = r_1 \cdot (m \cdot r_2) \quad (m \in M, r_1 \in R_1, r_2 \in R_2).$$

If R_1 and R_2 are unital rings and $1_{R_1} \cdot m = m = m \cdot 1_{R_2}$ for all $m \in M$, where 1_{R_1} is the unity of R_1 and 1_{R_2} is the unity of R_2, then we say that M is a *unital* (R_1, R_2)-*bimodule*. An (R, R)-bimodule is also called an *R-bimodule*.

For any rings R_1, R_2 and an (R_1, R_2)-bimodule M, the set of all matrices of the form

$$\begin{bmatrix} r_1 & m \\ 0 & r_2 \end{bmatrix},$$

where $r_1 \in R_1$, $m \in M$, and $r_2 \in R_2$, forms a ring under the usual matrix operations. We denote it by $\mathrm{Tri}(R_1, M, R_2)$ and call a *triangular ring*.

Corollary 2.5 *Let R_1 and R_2 be unital rings and let M be a unital (R_1, R_2)-bimodule. If R_1 and R_2 are generated by idempotents, then so is $R = \mathrm{Tri}(R_1, M, R_2)$ and is hence a zpd ring.*

Proof Every element in R can be written as

$$\begin{bmatrix} r_1 & 0 \\ 0 & r_2 \end{bmatrix} + \begin{bmatrix} 1_{R_1} & m \\ 0 & 0 \end{bmatrix} - \begin{bmatrix} 1_{R_1} & 0 \\ 0 & 0 \end{bmatrix}.$$

The first element lies in the subring generated by idempotents, while the other two elements are idempotents. $\square$

We now give a simple example of a unital zpd ring that is not generated by idempotents.

Example 2.6 Let R be the ring of all 2×2 matrices of the form

$$\begin{bmatrix} \alpha & \beta \\ \beta & \alpha \end{bmatrix} \quad (\alpha, \beta \in \mathbb{Z}).$$

It will be convenient to write such a matrix as $\alpha 1 + \beta u$ where 1 is the identity matrix and

$$u = \begin{bmatrix} 0 & 1 \\ 1 & 0 \end{bmatrix}.$$

We remark that $u^2 = 1$. One immediately checks that R has no nontrivial idempotents (i.e., idempotents different from 0 and 1). Now take a biadditive map $\varphi : R \times R \to R$ satisfying $\varphi(x, y) = 0$ whenever $xy = 0$. From

$$(1 - u)(1 + u) = 0 \quad \text{and} \quad (1 + u)(1 - u) = 0$$

it follows that

$$\varphi(1 - u, 1 + u) = 0 \quad \text{and} \quad \varphi(1 + u, 1 - u) = 0,$$

which further yields

$$\varphi(1, 1) = \varphi(u, u) \quad \text{and} \quad \varphi(1, u) = \varphi(u, 1).$$

Consequently,

$$\begin{aligned}
\varphi(\alpha 1 + \beta u, \gamma 1 + \delta u) &= (\alpha\gamma + \beta\delta)\varphi(1, 1) + (\alpha\delta + \beta\gamma)\varphi(u, 1) \\
&= \varphi\big((\alpha\gamma + \beta\delta)1 + (\alpha\delta + \beta\gamma)u, 1\big) \\
&= \varphi\big((\alpha 1 + \beta u)(\gamma 1 + \delta u), 1\big).
\end{aligned}$$

That is, $\varphi(x, y) = \varphi(xy, 1)$ holds for all $x, y \in R$, which proves that R is zpd.

2.2   Examples and Non-examples of zpd Algebras

We now return to the setting of algebras over fields. By the term "algebra", occurring without any adjective, we will always mean an associative algebra.

We start by showing that conditions (ii) and (iii) of Corollary 1.5 can be simplified if A is associative.

Proposition 2.7 *Let A be a unital algebra over a field F. The following conditions are equivalent:*

(i) *A is zpd.*
(ii) *If a bilinear functional $\varphi : A \times A \to F$ has the property that $\varphi(x, y) = 0$ whenever $x, y \in A$ are such that $xy = 0$, then*

$$\varphi(x, 1) = \varphi(1, x) \quad (x \in A).$$

(iii) *For all $w \in A$, $1 \otimes w - w \otimes 1 \in \text{span} \{x \otimes y \in A \otimes A \mid xy = 0\}$.*

Proof It is clear that (ii) follows from condition (ii) of Corollary 1.5. To show that the converse is true, take a bilinear $\varphi : A \times A \to F$ satisfying $\varphi(x, y) = 0$ whenever $xy = 0$, and introduce, for any $x \in A$, a new bilinear functional $\varphi_x : A \times A \to F$ by

$$\varphi_x(y, z) = \varphi(xy, z).$$

As A is associative, $yz = 0$ implies $(xy)z = 0$ and hence $\varphi_x(y, z) = 0$. Assuming (ii), it follows that $\varphi_x(y, 1) = \varphi_x(1, y)$ for all $y \in A$, i.e., $\varphi(xy, 1) = \varphi(x, y)$ for all $x, y \in A$. That is, A satisfies condition (ii) of Corollary 1.5.

Similarly, it is clear that (iii) follows from condition (iii) of Corollary 1.5, and we must show that the converse is also true. This is easy. Just write

$$1 \otimes w - w \otimes 1 = \sum_i x_i \otimes y_i$$

where $x_i y_i = 0$ for every i, and multiply this identity from the left by $z \otimes 1$ to obtain

$$z \otimes w - zw \otimes 1 = \sum_i zx_i \otimes y_i.$$

Note that $zx_i \cdot y_i = 0$ since A is associative. Thus, condition (iii) of Corollary 1.5 holds. $\square$

To state the next corollary, we have to recall some standard definitions. Most of them will be frequently used throughout the book. The first one was already mentioned in Example 1.10, but we repeat it due to its importance.

Throughout, A stands for an algebra.

Definition 2.8 A linear map $\delta : A \to A$ is called a *derivation* if it satisfies

$$\delta(xy) = \delta(x)y + x\delta(y) \quad (x, y \in A).$$

Let $x, y \in A$. Recall that the element

$$[x, y] = xy - yx$$

is called the *commutator* of x and y. Observe that for any $w \in A$, the map $x \mapsto [w, x]$ is a derivation. Such derivations have a special name.

Definition 2.9 A map $\delta : A \to A$ is called an *inner derivation* if there exists a $w \in A$ such that

$$\delta(x) = [w, x] \quad (x \in A).$$

Definition 2.10 For every $x \in A$, define $L_x, R_x : A \to A$ by

$$L_x(u) = xu \quad \text{and} \quad R_x(u) = ux.$$

We call L_x a *left multiplication operator* and R_x a *right multiplication operator*. Further, we call

$$M(A) = \left\{ \sum_i L_{x_i} R_{y_i} \mid x_i, y_i \in A \right\}$$

the *multiplication algebra* of A.

Note that for all $x, y \in A$,

$$L_x L_y = L_{xy}, \quad R_x R_y = R_{yx}$$

and

$$L_x R_y = R_y L_x,$$

from which we see that $M(A)$ is a subalgebra of the algebra of all linear operators on A. If a is unital, then $L_1 = R_1$ is the identity operator, and hence $M(A)$ contains every left multiplication operator L_x, every right multiplication operator R_x, and, accordingly, every inner derivation

$$\delta = L_w - R_w.$$

Definition 2.11 An algebra A is said to be *prime* if for any ideals I and J of A, $IJ = \{0\}$ implies $I = \{0\}$ or $J = \{0\}$.

Obvious examples are simple algebras. However, the class of prime algebras is much larger (in particular, it contains algebras without zero-divisors, but for obvious reasons these algebras are not interesting for us).

The next corollary will actually involve *centrally closed* prime algebras. Their definition is more technical and will not be needed elsewhere in the book. Therefore, we will not give it here and refer the reader to [47, Section 7.5]. Let us only mention that every unital simple ring, considered as an algebra over its center, is an example of such an algebra.

The following definition will be needed only in the next section, but it seems natural to state it after the definition of a prime algebra.

Definition 2.12 An algebra A is said to be *semiprime* if for any ideal I of A, $I^2 = \{0\}$ implies $I = \{0\}$.

Prime algebras are obviously semiprime, but not conversely. The direct product of two (or more) semiprime algebras is semiprime, but not prime.

We need just one more definition.

Definition 2.13 The *opposite algebra* of A, denoted A^{op}, is the algebra obtained by replacing the product in A by the product $\cdot$ defined by $x \cdot y = yx$.

We now have enough information to state and prove the following corollary, which shows that the zpd condition is intimately connected with the structure of inner derivations.

Corollary 2.14 *Let A be a unital algebra. If A is zpd, then for every inner derivation δ of A there exist x_i, $y_i \in A$, $i = 1, \ldots, n$, such that $\delta = \sum_{i=1}^{n} L_{x_i} R_{y_i}$ and $x_i y_i = 0$ for every i. If A is a centrally closed prime algebra, the converse implication is true as well.*

Proof Observe that

$$x \otimes y \mapsto L_x R_y$$

defines an epimorphism from the algebra $A \otimes A^{\mathrm{op}}$ onto the multiplication algebra $M(A)$. Moreover, this epimorphism is an isomorphism if A is a centrally closed prime algebra [47, Theorem 7.44]. Therefore, the corollary follows from the equivalence of conditions (i) and (iii) in Proposition 2.7. $\qquad\square$

In order to find examples of zpd algebras, we just repeat the arguments from the preceding section. Given any algebra A and a bilinear functional $\varphi : A \times A \to F$ satisfying $\varphi(x, y) = 0$ whenever $xy = 0$, the proof of Lemma 2.2 shows that

$$A_0 = \{r \in A \mid \varphi(xr, z) = \varphi(x, rz) \text{ for all } x, z \in A\}$$

is a subalgebra of A which contains all idempotents in A. Therefore, if A is generated by idempotents, then we have $\varphi(xy, z) = \varphi(x, yz)$ for all $x, y, z \in A$. If A is also unital, this yields $\varphi(xy, 1) = \varphi(x, y)$ for all $x, y \in A$. The following theorem therefore holds.

Theorem 2.15 *If a unital algebra A is generated by idempotents, then A is zpd.*

Remark 2.16 A slightly different proof of Theorem 2.15 can be given as follows. Let A be a unital algebra, let

$$U = \text{span} \{x \otimes y \in A \otimes A \mid xy = 0\}$$

and

$$A^0 = \{w \in A \mid 1 \otimes w - w \otimes 1 \in U\}.$$

Observe that

$$U(1 \otimes A) \subseteq U \quad \text{and} \quad (A \otimes 1)U \subseteq U.$$

Therefore,

$$1 \otimes wz - wz \otimes 1 = (1 \otimes w - w \otimes 1)(1 \otimes z) + (w \otimes 1)(1 \otimes z - z \otimes 1)$$

shows that A^0 is a subalgebra of A. Every idempotent $e \in A$ lies in A^0, as we see from

$$1 \otimes e - e \otimes 1 = (1 - e) \otimes e - e \otimes (1 - e).$$

Therefore, $A^0 = A$ if A is generated by idempotents. Note that, by Proposition 2.7, the condition that $A^0 = A$ means nothing but that A is zpd.

An advantage of this proof is that it shows explicitly how to express $1 \otimes w - w \otimes 1$, with w belonging to the subalgebra of A generated by idempotents, as an element in U. For example, if e_1 and e_2 are idempotents, then the above calculations show that

$$1 \otimes e_1 e_2 - e_1 e_2 \otimes 1 = (1 - e_1) \otimes e_1 e_2 - e_1 \otimes (1 - e_1)e_2$$
$$+ e_1(1 - e_2) \otimes e_2 - e_1 e_2 \otimes (1 - e_2).$$

From this we also see how to write inner derivations $\delta = L_w - R_w$ as operators of the form $\sum_{i=1}^{n} L_{x_i} R_{y_i}$ with $x_i y_i = 0$ for every i.

The algebra version of Corollaries 2.4 and 2.5 reads as follows.

Corollary 2.17 *The following algebras are zpd:*

(a) *The algebra $M_n(B)$, where $n \geq 2$ and B is any unital algebra.*
(b) *The algebra $\mathrm{Tri}(A_1, M, A_2)$, where A_1 and A_2 are unital algebras generated by idempotents and M is a unital (A_1, A_2)-bimodule.*

We point out the most typical examples of algebras of type (a) and (b).

Example 2.18 Let V be a vector space over a field F. Then $A = \mathrm{End}_F(V)$, the algebra of all linear operators from V to V, is of type (a). Indeed, if V is of finite dimension n, then $A \cong M_n(F)$, and if V is infinite-dimensional, then $A \cong M_n(A)$ for every positive integer n (see, for example, [47, Example 4.23]).

Example 2.19 The algebra $T_n(F)$ of all upper triangular $n \times n$ matrices over a field F is easily seen to be of type (b).

Remark 2.20 It is noteworthy that a simple unital algebra containing only one nontrivial idempotent is already generated by idempotents and is hence zpd. This follows from Lemma 2.26 below.

What about zpd algebras that are not generated by idempotents? Among algebras without unity it is easy to find such examples. Let us point out the simplest one.

Example 2.21 Any algebra A with trivial multiplication ($xy = 0$ for all $x, y \in A$) is zpd.

Among more interesting examples, we mention the algebra of strictly upper triangular matrices over a field. This was shown in [61] where one can find further examples. However, what is an example of a *unital* zpd algebra that is not generated by idempotents? This question is harder than it might first seem. In the next examples, we will examine some natural candidates for such an algebra.

Example 2.22 Perhaps the most obvious candidate is the algebra obtained by replacing the role of $\mathbb{Z}$ by a field F in Example 2.6. However, if the characteristic of F is not 2, then $\frac{1}{2}(1 + u)$ and $\frac{1}{2}(1 - u)$ are idempotents, so A is generated by idempotents in this case. On the other hand, if F has characteristic 2, then $(1 + u)^2 = 0$ and the only zero-divisors in A are scalar multiples of $1 + u$. Hence, Proposition 1.7 shows that A is not zpd.

Example 2.23 Another natural candidate may be the algebra obtained by adjoining a unity to a zpd algebra A without unity that is not generated by idempotents. However, Proposition 1.19 tells us that such an algebra is not zpd under a rather mild assumption that $A^2 \neq A$. Nevertheless, let us look more closely at the simplest case where A has trivial multiplication. Then $A^\#$, the algebra obtained by adjoining a unity to A, is a commutative unital algebra. We claim that $A^\#$ is *symmetrically* zpd. Take a symmetric bilinear functional φ on $A^\#$ that vanishes on pairs of elements with zero product. In particular, $\varphi(x, y) = 0$ for all $x, y \in A$. Hence,

$$\varphi(\alpha + x, \beta + y) = \alpha\beta\varphi(1, 1) + \alpha\varphi(1, y) + \beta\varphi(x, 1)$$

for all α, β in the corresponding field F and all $x, y \in A$. Since φ is symmetric, this yields

$$\varphi(\alpha + x, \beta + y) = \varphi(\alpha\beta, 1) + \varphi(\alpha y, 1) + \varphi(\beta x, 1)$$
$$= \varphi\big((\alpha + x)(\beta + y), 1\big).$$

Thus, φ is of the desired form, meaning that $A^\#$ is indeed an example of a symmetrically zpd algebra that is not zpd (by Proposition 1.19).

So, for example, the algebra $T_2(F)$ of all upper triangular 2×2 matrices is zpd (Example 2.19) and so is its subalgebra of all strictly upper triangular 2×2 matrices (which has trivial multiplication), while the algebra of all upper triangular matrices of the form

$$\begin{bmatrix} \alpha & \beta \\ 0 & \alpha \end{bmatrix} \quad (\alpha, \beta \in F)$$

is symmetrically zpd, but not zpd.

Example 2.24 We have proved that the direct sum of zpd algebras is zpd (Theorem 1.16). If the sum involves infinitely many terms, this algebra does not have a unity. It is therefore natural to look at the direct product of infinitely many unital zpd algebras. Let us consider the simplest possible example, that is, the direct product

$$A = F \times F \times \ldots$$

of countably infinitely many copies of the field F. If F is finite, then A is generated by idempotents. Assume, therefore, that F is infinite. We will prove that then A is not zpd.

We start the proof by taking $w = (\alpha_1, \alpha_2, \ldots) \in A$ such that $\alpha_i \neq \alpha_j$ for all $i \neq j$ (such exists since F is infinite). Assuming that A is zpd, we see from condition (iii) of Proposition 2.7 that there are $x_i, y_i \in A$ such that

$$1 \otimes w - w \otimes 1 = \sum_i x_i \otimes y_i$$

and $x_i y_i = 0$ for every i. Denote by X_i (resp. Y_i) the set of all $k \in \mathbb{N}$ such that the k-th term of x_i (resp. y_i) is 0. From $x_i y_i = 0$ we infer that $X_i \cup Y_i = \mathbb{N}$ for every i, and so

$$\mathbb{N} = (X_1 \cup Y_1) \cap (X_2 \cup Y_2) \cap \cdots \cap (X_n \cup Y_n).$$

Using the distributive law, it follows that

$$\mathbb{N} = \left(X_1 \cap \cdots \cap X_n\right) \cup \left(X_1 \cap \cdots \cap X_{n-1} \cap Y_n\right) \cup \cdots \cup \left(Y_1 \cap \cdots \cap Y_n\right).$$

Of course, whenever $\mathbb{N}$ is a union of finitely many sets, at least one of them must be infinite. Therefore, for each $i = 1, \ldots, n$, we can choose $W_i \in \{X_i, Y_i\}$ so that the set $W_1 \cap \cdots \cap W_n$ is infinite (we will actually only need that this set has at least two elements). Let I denote the ideal of A consisting of all $(\lambda_1, \lambda_2, \ldots)$ such that $\lambda_k = 0$ if $k \in W_1 \cap \cdots \cap W_n$. Note that $x_i \in I$ if $W_i = X_i$, and $y_i \in I$ if $W_i = Y_i$. Therefore,

$$w \otimes 1 - 1 \otimes w \in I \otimes A + A \otimes I,$$

meaning that there are $u_k, v_\ell \in I$ and $w_k, z_\ell \in A$ such that

$$w \otimes 1 - 1 \otimes w = \sum_{k=1}^{r} u_k \otimes w_k + \sum_{\ell=1}^{s} z_\ell \otimes v_\ell. \tag{2.1}$$

We claim that

$$I \cap \mathrm{span}\{w, 1\} = \{0\}. \tag{2.2}$$

Indeed, $1 \notin I$ and, unlike the elements in I, the elements of the form $\alpha w + \beta 1$ with $0 \neq \alpha, \beta \in F$ have mutually distinct terms. Without loss of generality, we may assume that $u_1, \ldots, u_r$ are linearly independent. Hence it follows from (2.2) that $w, 1, u_1, \ldots, u_r$ are linearly independent too. Now, applying Lemma 1.6 to (2.1) leads to a contradiction that $1 \in \mathrm{span}\{v_1, \ldots, v_s\} \subseteq I$. Thus, A is not zpd.

Some further examples of unital algebras that are not zpd are given in [1, 48, 78, 84]. All in all, we are forced to conclude that we do not know of any example of a unital zpd algebra that is not generated by idempotents. In the next section, we will show that such an example cannot be found among finite-dimensional algebras. We are, nevertheless, inclined to conjecture that it exists. Discovering it is a major problem since, without it, applications of the theory of associative zpd algebras have a limited reach.

2.3   The Finite-Dimensional Case

The goal of this section is to show that a finite-dimensional unital associative algebra is zpd if and only it is generated by idempotents. We start with two lemmas on idempotents that are interesting by themselves.

The first lemma can be extracted from the general theory of *lifting idempotents* modulo ideals [116, 137]. The proof we give was suggested by one of the referees of the book. It uses the following standard fact: If I is a minimal left ideal of an algebra A, then either $I^2 = \{0\}$ or $I = Af$ for some idempotent f [116, 10.22] (by a minimal left ideal we mean a nonzero left ideal that does not properly contain another nonzero left ideal).

Lemma 2.25 *Let I be an ideal of a finite-dimensional unital algebra A. Every idempotent in the quotient algebra A/I is of the form $e + I$ with e an idempotent in A.*

Proof Let $x \in A$ be such that $x + I$ is an idempotent in A/I. Denote by A_0 the (unital) subalgebra generated by x. As $x + A_0 \cap I$ is an idempotent in $A_0/A_0 \cap I$, it is enough to consider the case where $A = A_0$. In particular, we may thus assume that A is a commutative algebra. Also, we may assume that the lemma is true for all algebras of smaller dimension than A.

In the next step, we make the reduction to the case where I is minimal. We can assume that $I \neq \{0\}$, so I contains a minimal ideal J (which is simply a nonzero ideal of minimal dimension contained in I). Since the algebra A/J has lower dimension than A, we may use the above assumption (applied to the ideal I/J) to obtain an idempotent $u + J \in A/J$ such that $x + I = u + I$. Now, assuming the result is true for minimal ideals, it follows that there is an idempotent e in A such that $u + J = e + J$, and so $x + I = u + I = e + I$, as desired.

We may therefore assume that I is a minimal ideal. As mentioned before the statement of the lemma, we have either $I^2 = \{0\}$ or $I = Af$ for some idempotent f. In the latter case,

$$x + I = x(1 - f) + I$$

and $x(1 - f)$ is an idempotent since $x^2 - x \in I$ and A is commutative. Thus, let $I^2 = \{0\}$. Set

$$e = 3x^2 - 2x^3 = x - (2x - 1)(x^2 - x).$$

Since $x^2 - x \in I$, $x + I = e + I$. A straightforward computation shows that

$$e^2 - e = (x^2 - x)^2(4x^2 - 4x - 3) \in I^2 = \{0\},$$

so e is an idempotent.   □

The next lemma holds in general, not necessarily finite-dimensional unital algebras.

Lemma 2.26 *Let A be an algebra and let $e \in A$ be an idempotent. Then the ideal generated by all commutators $[e, x]$, $x \in A$, lies in the subalgebra generated by all idempotents in A.*

Proof By a straightforward calculation one verifies that every derivation δ of A satisfies

$$y\delta^3(x) = \delta\big(y\delta^2(x)\big) - \delta(y)\delta^2(x),$$

$$\delta^3(x)z = \delta\big(\delta^2(x)z\big) - \delta^2(x)\delta(z),$$

and

$$y\delta^3(x)z = \delta\big(y\delta^2(x)z\big) - \delta(y)\delta\big(\delta(x)z\big) - \delta\big(y\delta(x)\big)\delta(z) + 2\delta(y)\delta(x)\delta(z).$$

This shows that the ideal of A generated by the range of δ^3 is contained in the subalgebra of A generated by the range of δ. We need this in the special case where δ is the inner derivation $\delta(x) = [e, x]$. Note that $e^2 = e$ implies $\delta^3 = \delta$. Thus, this derivation has the property that the ideal and the subalgebra generated by its range coincide. A straightforward computation shows that

$$e + ex - exe \quad \text{and} \quad e + xe - exe$$

are idempotents. Since

$$\delta(x) = [e, x] = (e + ex - exe) - (e + xe - exe),$$

it follows that the ideal generated by all $\delta(x)$ is contained in the subalgebra generated by idempotents.   □

We now have enough information to prove the main result of the section. The proof is a nice illustration of the power of the classical Wedderburn's structure theory of finite-dimensional algebras, which we briefly survey in the next lines.

Let A be an algebra over a field F. We say that $a \in A$ is a *nilpotent element* if there exists a positive integer n such that $a^n = 0$. Similarly we define a *nilpotent ideal* as an ideal, I, such that $I^n = \{0\}$ for some $n \geq 1$. This means that $u_1 \cdots u_n = 0$ for all $u_i \in I$. A semiprime algebra can be equivalently defined as an algebra without nonzero nilpotent ideals.

Assume from now on that A is finite-dimensional. It is easy to see that A contains a unique maximal nilpotent ideal $\mathrm{rad}(A)$ which contains all other nilpotent ideals of A [47, Lemma 2.11]. It is called the (Jacobson) *radical* of A. Note that if A is unital, then every element $1 - r$ with $r \in \mathrm{rad}(A)$ is invertible. Indeed, if $r^n = 0$, then its inverse is $1 + r + \cdots + r^{n-1}$. Of course, $\mathrm{rad}(A) = \{0\}$ if and only if A is semiprime; a more common name for a finite-dimensional semiprime algebra is a *semisimple algebra*. We will need the following two well-known results:

(a) The algebra $A/\mathrm{rad}(A)$ is semisimple [47, Lemma 2.26].
(b) If A is semisimple, then there exist division algebras $D_1, \ldots, D_r$ over F such that

$$A \cong M_{n_1}(D_1) \times \cdots \times M_{n_r}(D_r) \tag{2.3}$$

for some positive integers $n_1, \ldots, n_r$ [47, Theorem 2.64].

Theorem 2.27 *A finite-dimensional unital algebra A is zpd if and only if it is generated by idempotents.*

Proof The "if" part has been established in Theorem 2.15. Thus, assume that A is zpd, and let us show that it is generated by idempotents. The proof will consist of four steps.

In the first step, we consider the situation where A is semisimple. We claim that then

$$A \cong M_{n_1}(D_1) \times \cdots \times M_{n_s}(D_s) \times F \times \cdots \times F, \tag{2.4}$$

where $n_i \geq 2$ and D_i are division algebras. Indeed, we know that A is isomorphic to an algebra of the form (2.3) described in (b). By Theorem 1.16, each of the algebras $A_i = M_{n_i}(D_i), i = 1, \ldots, r$, is zpd. If $n_i = 1$, then A_i has no zero-divisors and so Corollary 1.8 implies that $\dim_F A_i = 1$, i.e., $A_i \cong F$. This proves (2.4). Now, the algebras $M_{n_i}(D_i), = 1, \ldots, s$, are generated by idempotents by Corollary 2.4, while F is, of course, generated by the idempotent 1. Hence, A is generated by idempotents as well.

In the second step, we consider the case where A has no nontrivial idempotents. The algebra $A/\mathrm{rad}(A)$ is also zpd by Corollary 1.15. Moreover, it is semisimple (by (a)) and has no nontrivial idempotents by Lemma 2.25. From the first step it readily follows that $A/\mathrm{rad}(A) \cong F$. Thus, $\mathrm{rad}(A)$ has codimension 1 in A, so every element in A can be written as $\alpha 1 + r$ with $\alpha \in F$ and $r \in \mathrm{rad}(A)$. This element is invertible whenever $\alpha \neq 0$. The only zero-divisors in A are therefore the elements from $\mathrm{rad}(A)$. Since $1 \notin \mathrm{rad}(A)$, it follows from Proposition 1.7 that $\dim_F A = 1$ (and hence $\mathrm{rad}(A) = \{0\}$). Therefore,

$$A \cong F$$

in this case.

In the third step, we consider the case where all idempotents in A are central (i.e., they lie in the center of A). We claim that then

$$A \cong F \times \cdots \times F.$$

The proof is by induction on $n = \dim_F A$. The case where $n = 1$ is trivial, so let $n > 1$. From the conclusion of the second step it is evident that A must contain a (central) idempotent e different from 0 and 1. Note that eA and $(1 - e)A$ are (unital) algebras of smaller dimension than A and they both satisfy the condition that all of their idempotents are central. By the induction hypothesis, eA and $(1 - e)A$ can be represented as direct products of copies of F. But then the same is true for $A = eA \oplus (1 - e)A$.

We are ready for the final step of the proof. Let now A be arbitrary. Denote by A_0 the subalgebra of A generated by all idempotents. We must prove that $A_0 = A$. By the first step together with Corollary 1.15, every element $x + \mathrm{rad}(A) \in A/\mathrm{rad}(A)$ is a linear combination of products of idempotents in $A/\mathrm{rad}(A)$. Using Lemma 2.25, we thus see that there exists an $x_0 \in A_0$ such that $x + \mathrm{rad}(A) = x_0 + \mathrm{rad}(A)$. This means that

$$A = A_0 + \mathrm{rad}(A). \tag{2.5}$$

Let I denote the ideal generated by all commutators of idempotents with arbitrary elements in A. Since, by Lemma 2.25, all idempotents in A/I are of the form $e + I$ with e an idempotent in A, and since $[e, x] \in I$ for every $x \in A$, every idempotent in A/I is central. Corollary 1.15 tells us that A/I is zpd, and so

$$A/I \cong F \times \cdots \times F$$

by the third step. In particular, A/I is semisimple. Since $(I + \mathrm{rad}(A))/I$ is obviously a nilpotent ideal of A/I, it follows that $\mathrm{rad}(A) \subseteq I$. As $I \subseteq A_0$ by Lemma 2.26, we thus have $\mathrm{rad}(A) \subseteq A_0$, which together with (2.5) yields $A_0 = A$.   □

This theorem seems surprising since it is not obvious from the definitions that there should be any connection between zpd algebras and algebras generated by idempotents. On the other hand, its drawback is that it indicates that the class of zpd algebras is narrower than one might hope. As we already mentioned at the end of the preceding section, the question of the existence of infinite-dimensional unital zpd algebras that are not generated by idempotents is entirely open.

Most results and examples from this section are taken from [48]. Let us finally also mention the paper [100] which, by making use of the results of this section, provides another characterization of finite-dimensional unital algebras that are generated by idempotents.

Zero Lie/Jordan Product Determined Algebras

In this chapter, we will be primarily interested in the question of which associative algebras, viewed either as Lie algebras under the Lie product or as Jordan algebras under the Jordan product, are zero product determined. We will see that the results in the Jordan case are similar to those in the associative case, while the Lie case is quite different. For example, a unital associative algebra (over a field of characteristic not 2) that is generated by idempotents is zpd when considered as a Jordan algebra, but not necessarily when considered as a Lie algebra. On the other hand, it is easy to find examples of associative algebras that are zpd as Lie algebras, but not as associative algebras.

3.1   Lie Algebras and Jordan Algebras

We start with the definition of a Lie algebra.

Definition 3.1 A *Lie algebra* is a nonassociative algebra L whose multiplication, which we denote by $[\,\cdot\,,\,\cdot\,]$, satisfies the following axioms:

$$[x, x] = 0 \quad \text{and} \quad [[x, y], z] + [[z, x], y] + [[y, z], x] = 0.$$

The second axiom is called the *Jacobi identity*.

Every associative algebra A turns into a Lie algebra, denoted A^-, by replacing the original product by the *Lie product*

$$[x, y] = xy - yx.$$

M. Brešar, *Zero Product Determined Algebras*, Frontiers in Mathematics, https://doi.org/10.1007/978-3-030-80242-4_3

Thus, the Lie product of two elements is nothing but their commutator and is equal to zero if and only if the elements commute.

We will be primarily interested in the question of which Lie algebras of type A^- are zpd. This is the theme of Sect. 3.2.

Let us say here a few words on Lie algebras of general type. The question of which of them are zpd is interesting and has gained some attention [60, 87, 165] but requires techniques that are different from those used elsewhere in this book. Presenting proofs would thus take us too far afield, so we only give some general comments. Given a concrete Lie algebra, it is usually not easy to find out (and even not easy to conjecture) whether or not it is zpd. No general method to attack this question has been developed so far. It is noteworthy that the finite-dimensional semisimple complex Lie algebras are zpd [165], and so are all Lie algebras of dimension less than four [60]. On the other hand, [60] provides examples of several important Lie algebras that are not zpd. Generally speaking, the problem of determining which Lie algebras are and which are not zpd is largely open.

We add three general remarks concerning zpd Lie algebras.

Remark 3.2 The concept of a zpd Lie algebra is analogous to the group-theoretic notion of triviality of Bogomolov multiplier (see, for example, [135]).

Remark 3.3 It is noteworthy to mention an auxiliary notion introduced in [60]. A module V over a Lie algebra L over a field F is said to be *zero action determined* (*zad* for short) if, for every bilinear functional $\varphi : L \times V \to F$ with the property that for all $x \in L$, $v \in V$,

$$xv = 0 \Longrightarrow \varphi(x, v) = 0,$$

there exists a linear functional τ on LV such that

$$\varphi(x, v) = \tau(xv)$$

for all $x \in L$, $v \in V$. Observe that L is a zpd Lie algebra if and only if its adjoint module is zad. Thus, the zad notion may be viewed as a generalization of the zpd notion. In [60], however, it was used only as a tool for showing that some Lie algebras are zpd.

In essentially the same way, one can define zad modules over associative algebras. It is easy to see that a unital associative algebra A is zpd if and only if every unital A-module V is zad [60, 100]. This is not true for Lie algebras [60, Remark 2.11].

We will not use zad modules in what follows. We have mentioned this concept primarily because it could be interesting for module-theoretic oriented mathematicians.

Remark 3.4 The definition of a zpd Lie algebra may remind one of the notion of the second cohomology group of a Lie algebra. Let us briefly record the corresponding definitions. A bilinear functional φ on a Lie algebra L over a field F is a called a 2-*cocycle* if it is *skew-symmetric*, i.e.,

$$\varphi(x, y) = -\varphi(y, x) \quad (x, y \in L),$$

and satisfies

$$\varphi([x, y], z) + \varphi([z, x], y) + \varphi([y, z], x) = 0 \quad (x, y, z \in L).$$

An example of a 2-cocycle is any *coboundary*. This is a bilinear functional φ of the form

$$\varphi(x, y) = \tau([x, y]),$$

where τ is a linear functional on L. The condition that every 2-cocycle on L is a coboundary can be stated as that the *second cohomology group* $H^2(L, F)$ is trivial. On the other hand, the condition that L is zpd can be stated as that every bilinear functional φ on L with the property that $[x, y] = 0$ implies $\varphi(x, y) = 0$ is a coboundary. The problem of showing that a Lie algebra L is zpd is thus obviously similar to the problem of showing that the second cohomology group $H^2(L, F)$ is trivial; however, the two problems are, in general, independent. In terms of tensor products, the zpd condition can be, by Proposition 1.3, presented as the equality of the linear spaces

$$V = \left\{ \sum_i z_i \otimes w_i \mid \sum_i [z_i, w_i] = 0 \right\}$$

and

$$U = \text{span} \{ x \otimes y \mid x, y \in L, \ [x, y] = 0 \}.$$

One can similarly check that the triviality of $H^2(L, F)$ is equivalent to the condition that V is equal to

$$U' = \text{span} \{ x \otimes y + y \otimes x, [x, y] \otimes z + [z, x] \otimes y + [y, z] \otimes x \mid x, y, z \in L \}.$$

Unfortunately, there does not seem to exist some simple connection between the spaces U and U'.

Let us finally briefly introduce the second important type of nonassociative algebras with which we will deal with.

Definition 3.5 A *Jordan algebra* is a nonassociative algebra J whose multiplication, which we denote by $\circ$, satisfies the following axioms:

$$x \circ y = y \circ x \quad \text{and} \quad ((x \circ x) \circ y) \circ x = (x \circ x) \circ (y \circ x).$$

Every associative algebra A becomes a Jordan algebra, denoted A^+, if we replace the original product by the *Jordan product*

$$x \circ y = xy + yx.$$

Section 3.3 is devoted to the question of when these Jordan algebras are zpd. There have been some attempts to consider this question for Jordan algebras of other types [88, 89], but we shall not go into this here.

3.2   zLpd Algebras

The theme of this section is the following notion.

Definition 3.6 An associative algebra A is *zero Lie product determined* if the Lie algebra A^- is zero product determined.

Throughout, we will write *zLpd* as short for "zero Lie product determined." Thus, A is zLpd if, for every bilinear functional φ on A satisfying $\varphi(x, y) = 0$ whenever x and y commute, there exists a linear functional τ on A such that $\varphi(x, y) = \tau([x, y])$ for all $x, y \in A$.

Let us first record the most obvious example, which in particular shows that not every zLpd algebra is zpd.

Example 3.7 Every commutative algebra is zLpd.

It is more difficult to find algebras that are not zLpd. We now give an example (taken from [18]) of such an algebra which is, moreover, finite-dimensional. It will be obtained by adding another relation to those defining the Grassmann algebra in four generators.

Example 3.8 Let F be a field of characteristic not 2. Let A be the unital algebra over F generated by four elements y_1, y_2, y_3, y_4 and relations

$$y_1 y_2 = y_3 y_4, \quad y_i^2 = 0, \quad \text{and} \quad y_i y_j = -y_j y_i$$

for all $i, j \in \{1, 2, 3, 4\}$. That is, A is the free algebra $F\langle x_1, x_2, x_3, x_4 \rangle$ in four generators x_1, x_2, x_3, x_4 modulo the ideal J generated by $x_1 x_2 - x_3 x_4$, x_i^2, $x_i x_j + x_j x_i$ (so $y_i = x_i + J$).
One can verify that the elements

$$1, \ y_1, \ y_2, \ y_3, \ y_4, \ y_1 y_2, \ y_1 y_3, \ y_1 y_4, \ y_2 y_3, \ y_2 y_4$$

form a basis of A. Therefore, A is 10-dimensional. Observe that each $y_i y_j$ commutes with any element from our basis, so it belongs to the center $Z(A)$ of A. Every element in A can therefore be written as a sum of an element in $Z(A)$ and an element in span $\{y_1, y_2, y_3, y_4\}$.

Define the bilinear functional $\varphi \colon A \times A \to F$ by

$$\varphi(y_1, y_2) = -\varphi(y_2, y_1) = 1$$

and

$$\varphi(b, b') = 0$$

for all other pairs of elements from our basis. Pick $x, y \in A$ and write

$$x = \alpha_1 y_1 + \alpha_2 y_2 + \alpha_3 y_3 + \alpha_4 y_4 + z \quad \text{and} \quad y = \beta_1 y_1 + \beta_2 y_2 + \beta_3 y_3 + \beta_4 y_4 + w,$$

with $\alpha_i, \beta_j \in F$ and $z, w \in Z(A)$. Suppose x and y commute. Then

$$\left[\alpha_1 y_1 + \alpha_2 y_2 + \alpha_3 y_3 + \alpha_4 y_4, \beta_1 y_1 + \beta_2 y_2 + \beta_3 y_3 + \beta_4 y_4\right] = 0,$$

which gives

$$\big((\alpha_1\beta_2 - \alpha_2\beta_1) + (\alpha_3\beta_4 - \alpha_4\beta_3)\big) y_1 y_2$$
$$+ (\alpha_1\beta_3 - \alpha_3\beta_1) y_1 y_3 + (\alpha_2\beta_3 - \alpha_3\beta_2) y_2 y_3$$
$$+ (\alpha_1\beta_4 - \alpha_4\beta_1) y_1 y_4 + (\alpha_2\beta_4 - \alpha_4\beta_2) y_2 y_4 = 0$$

and hence

$$(\alpha_1\beta_2 - \alpha_2\beta_1) + (\alpha_3\beta_4 - \alpha_4\beta_3) = 0, \tag{3.1}$$

$$\alpha_1\beta_3 = \alpha_3\beta_1, \quad \alpha_2\beta_3 = \alpha_3\beta_2, \tag{3.2}$$

$$\alpha_1\beta_4 = \alpha_4\beta_1, \quad \alpha_2\beta_4 = \alpha_4\beta_2. \tag{3.3}$$

From (3.2) we derive

$$(\alpha_1\beta_2 - \alpha_2\beta_1)\beta_3 = 0.$$

Similarly, from (3.3) we derive

$$(\alpha_1\beta_2 - \alpha_2\beta_1)\beta_4 = 0.$$

Along with (3.1), these identities readily imply

$$\varphi(x, y) = \alpha_1\beta_2 - \alpha_2\beta_1 = 0.$$

We have shown that φ vanishes on pairs of commuting elements. However,

$$\varphi(y_1, y_2) \neq \varphi(y_3, y_4)$$

while $[y_1, y_2] = [y_3, y_4]$, so φ is not of the form $\varphi(x, y) = \tau([x, y])$. This means that A is not zLpd.

We continue with a more sophisticated example, which shows more than just that the algebra under consideration is not zLpd. This additional information will be needed later, in Example 3.13.

Example 3.9 Let F be a field, let $n \geq 2$, and let $A = F\langle x_1, x_2, \ldots, x_{2n}\rangle$ be the free algebra in $x_1, \ldots, x_{2n}$. Denote by I the ideal of A generated by the polynomial

$$[x_1, x_2] + [x_3, x_4] + \ldots + [x_{2n-1}, x_{2n}].$$

Set $B_n = A/I$. Thus, B_n is the (unital) algebra generated by $2n$ elements $u_i = x_i + I$, $i = 1, \ldots, 2n$, and the relation

$$[u_1, u_2] + [u_3, u_4] + \ldots + [u_{2n-1}, u_{2n}] = 0. \tag{3.4}$$

It is clear that $u_1, \ldots, u_{2n}$ are linearly independent. Take a basis U of B_n that contains $u_1, \ldots, u_{2n}$ and define the bilinear functional φ on B_n by

$$\varphi(u_1, u_2) = -\varphi(u_2, u_1) = 1$$

and

$$\varphi(u, u') = 0$$

for all other pairs of elements $u, u' \in U$. Since

$$\varphi(u_1, u_2) + \varphi(u_3, u_4) + \cdots + \varphi(u_{2n-1}, u_{2n}) = 1, \tag{3.5}$$

it follows from (3.4) that φ is not of the form $\varphi(x, y) = \tau([x, y])$.

Our goal is to show that for all $v_t, w_t \in B_n, t = 1, \ldots, n-1$,

$$\sum_{t=1}^{n-1} [v_t, w_t] = 0 \implies \sum_{t=1}^{n-1} \varphi(v_t, w_t) = 0. \tag{3.6}$$

We start the proof of (3.6) by writing

$$v_t = \lambda_t u_1 + \mu_t u_2 + p_t \quad \text{and} \quad w_t = \alpha_t u_1 + \beta_t u_2 + q_t,$$

where $\lambda_t, \mu_t, \alpha_t, \beta_t \in F$ and p_t, q_t lie in the linear span of $U \setminus \{u_1, u_2\}$. Observe that

$$\sum_{t=1}^{n-1} \varphi(v_t, w_t) = \sum_{t=1}^{n-1} (\lambda_t \beta_t - \mu_t \alpha_t).$$

Thus, our goal is to show that

$$\sum_{t=1}^{n-1} [v_t, w_t] = 0 \implies \sum_{t=1}^{n-1} (\lambda_t \beta_t - \mu_t \alpha_t) = 0. \tag{3.7}$$

We can write

$$p_t = r_t + f_t + I \quad \text{and} \quad q_t = s_t + g_t + I,$$

where $r_t, s_t \in \text{span}\{x_3, \ldots, x_{2n}\}$ and f_t, g_t lie in the linear span of monomials of degree 0 or at least 2. From $\sum_{t=1}^{n-1} [v_t, w_t] = 0$ we infer that

$$\sum_{t=1}^{n-1} [\lambda_t x_1 + \mu_t x_2 + r_t + f_t, \alpha_t x_1 + \beta_t x_2 + s_t + g_t] \in I.$$

Thus,

$$\sum_{t=1}^{n-1} [\lambda_t x_1 + \mu_t x_2 + r_t + f_t, \alpha_t x_1 + \beta_t x_2 + s_t + g_t]$$

$$= \gamma \left([x_1, x_2] + [x_3, x_4] + \ldots + [x_{2n-1}, x_{2n}] \right) + h,$$

where $\gamma \in F$ and $h \in I$ is a linear combination of monomials of degree at least 3. Taking into account the degrees of the polynomials in this identity it follows that

$$\sum_{t=1}^{n-1} [\lambda_t x_1 + \mu_t x_2 + r_t, \alpha_t x_1 + \beta_t x_2 + s_t]$$

$$= \gamma \Big([x_1, x_2] + [x_3, x_4] + \ldots + [x_{2n-1}, x_{2n}] \Big).$$

Assuming that $\gamma \neq 0$, we have that $[x_1, x_2] + [x_3, x_4] + \ldots + [x_{2n-1}, x_{2n}]$ can be written as a sum of $n - 1$ commutators in A. However, this is impossible. To show this, assume that $a_i, b_i \in A, i = 1, \ldots, m$, are such that

$$[a_1, b_1] + [a_2, b_2] + \ldots + [a_m, b_m] = [x_1, x_2] + [x_3, x_4] + \ldots + [x_{2n-1}, x_{2n}]. \qquad (3.8)$$

Let us prove that then $m \geq n$. We proceed by induction on n. The $n = 1$ case is trivial, so let $n > 1$. By a degree consideration, we may assume that all a_i and b_i lie in span $\{x_1, \ldots, x_{2n}\}$. In particular,

$$b_m = \mu_1 x_1 + \cdots + \mu_{2n-1} x_{2n-1} + \mu_{2n} x_{2n}$$

for some $\mu_j \in F$. We may assume that $\mu_{2n} \neq 0$. Since (3.8) is an identity in a free algebra, we may replace the indeterminate x_i by any element in A. Now, replacing x_{2n-1} by 0 and x_{2n} by $-\sum_{j=1}^{2n-2} \mu_{2n}^{-1} \mu_j x_j$ gives

$$[c_1, d_1] + \ldots + [c_{m-1}, d_{m-1}] = [x_1, x_2] + \ldots + [x_{2n-3}, x_{2n-2}]$$

for some $c_i, d_i \in$ span $\{x_1, \ldots, x_{2n-2}\}$. We can now use the induction assumption to obtain $m - 1 \geq n - 1$ and hence $m \geq n$, as desired.

Therefore, $\gamma = 0$, and hence

$$0 = \sum_{t=1}^{n-1} [\lambda_t x_1 + \mu_t x_2 + r_t, \alpha_t x_1 + \beta_t x_2 + s_t]$$

$$= \Big(\sum_{t=1}^{n-1} (\lambda_t \beta_t - \mu_t \alpha_t) \Big) [x_1, x_2] + k,$$

where k lies in the linear span of monomials different from $x_1 x_2$ and $x_2 x_1$. This yields

$$\sum_{t=1}^{n-1} (\lambda_t \beta_t - \mu_t \alpha_t) = 0,$$

proving (3.7).

Let us summarize what we have shown: For each $n \geq 2$, there exists a unital algebra B_n having a bilinear functional φ with property (3.6), but containing elements $u_1, \ldots, u_{2n}$ satisfying (3.4) and (3.5). In particular, B_n therefore is not zLpd.

In one of the pioneering papers on zpd algebras, [53], it was shown that, using the present terminology, the matrix algebra $M_n(F)$, where $n \geq 2$ and F is any field, is zLpd. This result was then used as an essential tool for describing linear maps from $M_n(F)$ to itself that preserve commutativity, in the sense that they send commuting pairs of elements into commuting pairs—see Theorem 7.14 below. One may ask whether, more generally, the algebra $M_n(B)$, where $n \geq 2$ and B is any unital algebra, is zLpd. This question is natural since such an algebra is always zpd (Corollary 2.17). We will show, however, that the answer is negative in general, but positive under the assumption that B is zLpd. These results were established in [58]. The rest of this section is devoted to their proofs.

We start with a positive result.

Theorem 3.10 *If B is a zLpd (associative) unital algebra, then so is the matrix algebra $M_n(B)$ for every $n \geq 2$.*

Proof We write A for $M_n(B)$. Let $\varphi : A \times A \to F$ be a bilinear functional such that for all $x, y \in A$, $[x, y] = 0$ implies $\varphi(x, y) = 0$. In particular, $\varphi(x, x) = 0$. Replacing x by $x + y$ we see that

$$\varphi(x, y) = -\varphi(y, x) \quad (x, y \in A),$$

i.e., φ is skew-symmetric. In what follows, this will be used without explicit mention.

Since x commutes with x^2, we have $\varphi(x^2, x) = 0$. Replacing x by $x + y$ we obtain

$$\varphi(x \circ y, x) + \varphi(x^2, y) + \varphi(x \circ y, y) + \varphi(y^2, x) = 0,$$

where, as before, $x \circ y$ stands for $xy + yx$. Now, replacing y by $y + z$ in the last identity results in

$$\varphi(x \circ y, z) + \varphi(z \circ x, y) + \varphi(y \circ z, x) = 0 \quad (x, y, z \in A). \tag{3.9}$$

Let e_{ij} and be_{ij} have the usual meaning (as in the proof of Corollary 2.4). We will now derive several identities involving these elements. Throughout, by a and b we denote arbitrary elements in B, and by i, j, k, l arbitrary elements in $\{1, \ldots, n\}$. Note that

$$ae_{ij} \cdot be_{kl} = \delta_{jk} abe_{il},$$

where δ_{jk} is the Kronecker delta.

Since ae_{ii} commutes with e_{ii}, we have

$$\varphi(ae_{ii}, e_{ii}) = 0. \tag{3.10}$$

Similarly,

$$\varphi(ae_{ij}, be_{kl}) = 0 \quad \text{if } j \neq k \text{ and } i \neq l. \tag{3.11}$$

Assume now that $i \neq j$. As $ae_{ij} + ae_{ji}$ commutes with $e_{ij} + e_{ji}$, we have

$$\varphi(ae_{ij} + ae_{ji}, e_{ij} + e_{ji}) = 0.$$

Since, by (3.11), $\varphi(ae_{ij}, e_{ij}) = 0$ and $\varphi(ae_{ji}, e_{ji}) = 0$, it follows that

$$\varphi(ae_{ij}, e_{ji}) = -\varphi(ae_{ji}, e_{ij}) \quad \text{if } i \neq j. \tag{3.12}$$

Our next goal is to prove

$$\varphi(ae_{ij}, be_{jk}) = \varphi(abe_{ik}, e_{kk}) = -\varphi(abe_{ik}, e_{ii}) \quad \text{if } i \neq k. \tag{3.13}$$

Note first that $[abe_{ik}, e_{ii} + e_{kk}] = 0$ and so $\varphi(abe_{ik}, e_{ii} + e_{kk}) = 0$, i.e.,

$$\varphi(abe_{ik}, e_{kk}) = -\varphi(abe_{ik}, e_{ii}).$$

To complete the proof of (3.13), we consider two cases: $j \neq k$ and $j = k$. In the first case, $ae_{ij} + abe_{ik}$ commutes with $be_{jk} - e_{kk}$ and therefore

$$\varphi(ae_{ij} + abe_{ik}, be_{jk} - e_{kk}) = 0.$$

Since $\varphi(ae_{ij}, e_{kk}) = 0$ and $\varphi(abe_{ik}, be_{jk}) = 0$ by (3.11), the desired identity

$$\varphi(ae_{ij}, be_{jk}) = \varphi(abe_{ik}, e_{kk})$$

follows. Consider now the second case where $j = k$. As $ae_{ik} - e_{ii}$ and $abe_{ik} + be_{kk}$ commute,

$$\varphi(ae_{ik} - e_{ii}, abe_{ik} + be_{kk}) = 0.$$

According to (3.11), $\varphi(ae_{ik}, abe_{ik}) = 0$ and $\varphi(e_{ii}, be_{kk}) = 0$, so we obtain

$$\varphi(ae_{ik}, be_{kk}) = \varphi(e_{ii}, abe_{ik}).$$

Since φ is skew-symmetric, we can write this as

$$\varphi(ae_{ik}, be_{kk}) = -\varphi(abe_{ik}, e_{ii}),$$

which completes the proof of (3.13).

Next, let us prove that

$$\varphi(ae_{ij}, be_{ji}) = \varphi(abe_{ij}, e_{ji}) + \varphi(ae_{jj}, be_{jj}). \tag{3.14}$$

If $i = j$, then this follows from (3.10). If $i \neq j$, then, using (3.9), we have

$$\varphi(ae_{ij}, be_{ji}) = \varphi(e_{ij} \circ ae_{jj}, be_{ji})$$
$$= -\varphi(be_{ji} \circ e_{ij}, ae_{jj}) - \varphi(ae_{jj} \circ be_{ji}, e_{ij})$$
$$= -\varphi(be_{jj} + be_{ii}, ae_{jj}) - \varphi(abe_{ji}, , e_{ij}).$$

Since $\varphi(be_{ii}, ae_{jj}) = 0$ by (3.11) and $\varphi(abe_{ji}, e_{ij}) = -\varphi(abe_{ij}, e_{ji})$ by (3.12) and (3.14) follows.

Finally, let us prove that

$$\varphi(ae_{ij}, be_{ji}) = \varphi(abe_{ik}, e_{ki}) - \varphi(bae_{jk}, e_{kj}) + \varphi(ae_{kk}, be_{kk}). \tag{3.15}$$

Assume first that $i \neq j$. If $k = j$, then (3.15) follows from (3.10) and (3.14). Using the skew-symmetry of φ, we see that the same is true if $k = i$. If $k \neq i$ and $k \neq j$, then we can use (3.9) to obtain

$$\varphi(ae_{ij}, be_{ji}) = \varphi(ae_{ik} \circ e_{kj}, be_{ji})$$
$$= -\varphi(be_{ji} \circ ae_{ik}, e_{kj}) - \varphi(e_{kj} \circ be_{ji}, ae_{ik})$$
$$= -\varphi(bae_{jk}, e_{kj}) + \varphi(ae_{ik}, be_{ki}),$$

and so (3.15) follows from (3.14). Now assume that $i = j$. In light of (3.10), we may assume $k \neq i$. Applying (3.9), we easily derive

$$\varphi(ae_{ii}, be_{ii}) = \varphi(ae_{ik} \circ e_{ki}, be_{ii})$$
$$= -\varphi(bae_{ik}, e_{ki}) + \varphi(ae_{ik}, be_{ki}).$$

Along with (3.14), this yields (3.15).

Now take any $x_t, y_t \in A$ such that $\sum_{t=1}^{m}[x_t, y_t] = 0$. According to Proposition 1.3 (ii), we have to prove that $\sum_{t=1}^{m} \varphi(x_t, y_t) = 0$. Write

$$x_t = \sum_{i=1}^{n} \sum_{j=1}^{n} a_{ij}^{t} e_{ij} \quad \text{and} \quad y_t = \sum_{k=1}^{n} \sum_{l=1}^{n} b_{kl}^{t} e_{kl},$$

where $a_{ij}^{t}, b_{kl}^{t} \in B$. Computing the (i, l) entry of $[x_t, y_t]$, we see that

$$\sum_{t=1}^{m} \sum_{j=1}^{n} (a_{ij}^{t} b_{jl}^{t} - b_{ij}^{t} a_{jl}^{t}) = 0 \text{ for all } i, l. \tag{3.16}$$

We start the computation of $\sum_{t=1}^{m} \varphi(x_t, y_t)$ by applying (3.11):

$$\sum_{t=1}^{m} \varphi(x_t, y_t) = \sum_{t=1}^{m} \sum_{i=1}^{n} \sum_{j=1}^{n} \sum_{k=1}^{n} \sum_{l=1}^{n} \varphi(a_{ij}^{t} e_{ij}, b_{kl}^{t} e_{kl})$$

$$= \sum_{t=1}^{m} \sum_{i=1}^{n} \sum_{j=1}^{n} \sum_{\substack{l=1 \\ l \neq i}}^{n} \varphi(a_{ij}^{t} e_{ij}, b_{jl}^{t} e_{jl})$$

$$+ \sum_{t=1}^{m} \sum_{i=1}^{n} \sum_{j=1}^{n} \sum_{\substack{k=1 \\ k \neq j}}^{n} \varphi(a_{ij}^{t} e_{ij}, b_{ki}^{t} e_{ki})$$

$$+ \sum_{t=1}^{m} \sum_{i=1}^{n} \sum_{j=1}^{n} \varphi(a_{ij}^{t} e_{ij}, b_{ji}^{t} e_{ji}).$$

For convenience, we rewrite the second summation as

$$\sum_{t=1}^{m} \sum_{i=1}^{n} \sum_{j=1}^{n} \sum_{\substack{l=1 \\ l \neq i}}^{n} \varphi(a_{jl}^{t} e_{jl}, b_{ij}^{t} e_{ij}) = -\sum_{t=1}^{m} \sum_{i=1}^{n} \sum_{j=1}^{n} \sum_{\substack{l=1 \\ l \neq i}}^{n} \varphi(b_{ij}^{t} e_{ij}, a_{jl}^{t} e_{jl}).$$

Applying (3.13) we now see that the sum of the first and the second summation is equal to

$$\sum_{t=1}^{m} \sum_{i=1}^{n} \sum_{j=1}^{n} \sum_{\substack{l=1 \\ l \neq i}}^{n} \left(\varphi(a_{ij}^{t} b_{jl}^{t} e_{il}, e_{ll}) - \varphi(b_{ij}^{t} a_{jl}^{t} e_{il}, e_{ll}) \right)$$

$$= \sum_{t=1}^{m} \sum_{i=1}^{n} \sum_{j=1}^{n} \sum_{\substack{l=1 \\ l \neq i}}^{n} \varphi((a_{ij}^{t} b_{jl}^{t} - b_{ij}^{t} a_{jl}^{t}) e_{il}, e_{ll})$$

$$= \sum_{i=1}^{n} \sum_{\substack{l=1 \\ l \neq i}}^{n} \varphi\left(\left(\sum_{t=1}^{m} \sum_{j=1}^{n} (a_{ij}^{t} b_{jl}^{t} - b_{ij}^{t} a_{jl}^{t}) \right) e_{il}, e_{ll} \right) = 0$$

by (3.16). Hence,

$$\sum_{t=1}^{m} \varphi(x_t, y_t) = \sum_{t=1}^{m}\sum_{i=1}^{n}\sum_{j=1}^{n} \varphi(a_{ij}^t e_{ij}, b_{ji}^t e_{ji}).$$

We have to prove that this summation also equals zero. Applying (3.15), we have

$$\varphi(a_{ij}^t e_{ij}, b_{ji}^t e_{ji}) = \varphi(a_{ij}^t b_{ji}^t e_{i1}, e_{1i}) - \varphi(b_{ji}^t a_{ij}^t e_{j1}, e_{1j}) + \varphi(a_{ij}^t e_{11}, b_{ji}^t e_{11}).$$

Thus,

$$\sum_{t=1}^{m} \varphi(x_t, y_t) = \sum_{t=1}^{m}\sum_{i=1}^{n}\sum_{j=1}^{n} \varphi(a_{ij}^t b_{ji}^t e_{i1}, e_{1i})$$

$$- \sum_{t=1}^{m}\sum_{i=1}^{n}\sum_{j=1}^{n} \varphi(b_{ji}^t a_{ij}^t e_{j1}, e_{1j})$$

$$+ \sum_{t=1}^{m}\sum_{i=1}^{n}\sum_{j=1}^{n} \varphi(a_{ij}^t e_{11}, b_{ji}^t e_{11}).$$

Rewriting, for convenience, the second summation as

$$- \sum_{t=1}^{m}\sum_{i=1}^{n}\sum_{j=1}^{n} \varphi(b_{ij}^t a_{ji}^t e_{i1}, e_{1i}),$$

we have

$$\sum_{t=1}^{m} \varphi(x_t, y_t) = \sum_{t=1}^{m}\sum_{i=1}^{n}\sum_{j=1}^{n} \varphi((a_{ij}^t b_{ji}^t - b_{ij}^t a_{ji}^t)e_{i1}, e_{1i}) + \sum_{t=1}^{m}\sum_{i=1}^{n}\sum_{j=1}^{n} \varphi(a_{ij}^t e_{11}, b_{ji}^t e_{11}).$$

The first summation can be written as

$$\sum_{i=1}^{n} \varphi\left(\left(\sum_{t=1}^{m}\sum_{j=1}^{n}(a_{ij}^t b_{ji}^t - b_{ij}^t a_{ji}^t)\right) e_{i1}, e_{1i}\right),$$

and so it is equal to 0 by (3.16). The proof will be completed by showing that the second summation also equals 0. That is, we must prove that

$$\sum_{t=1}^{m}\sum_{i=1}^{n}\sum_{j=1}^{n} \varphi(a_{ij}^t e_{11}, b_{ji}^t e_{11}) = 0. \tag{3.17}$$

To this end, consider the bilinear functional ψ on B defined by

$$\psi(a, b) = \varphi(ae_{11}, be_{11}).$$

Obviously, $[a, b] = 0$ implies $[ae_{11}, be_{11}] = 0$ and hence $\psi(a, b) = 0$. Since B is zLpd by assumption, we have

$$\sum_{t=1}^{m}[a_t, b_t] = 0 \implies \sum_{t=1}^{m}\psi(a_t, b_t) = 0. \tag{3.18}$$

As a special case of (3.16) we have

$$\sum_{t=1}^{m}\sum_{j=1}^{n}(a_{ij}^t b_{ji}^t - b_{ij}^t a_{ji}^t) = 0$$

for every i. Hence,

$$\sum_{t=1}^{m}\sum_{i=1}^{n}\sum_{j=1}^{n}(a_{ij}^t b_{ji}^t - b_{ij}^t a_{ji}^t) = 0,$$

which can be, after changing the roles of i and j in the second term, written as

$$\sum_{t=1}^{m}\sum_{i=1}^{n}\sum_{j=1}^{n}[a_{ij}^t, b_{ji}^t] = 0.$$

Applying (3.18) we obtain

$$\sum_{t=1}^{m}\sum_{i=1}^{n}\sum_{j=1}^{n}\psi(a_{ij}^t, b_{ji}^t) = 0,$$

which is just another form of writing (3.17).   □

Corollary 3.11 *If B is a commutative (associative) unital algebra, then $M_n(B)$ is a zLpd algebra for every $n \geq 2$.*

Proof Commutative algebras are trivially zLpd.   □

Remark 3.12 Let B be a zLpd algebra. Since $M_n(F)$ is zLpd, Theorem 1.18 implies that the algebra $M_n(F)^- \otimes B^-$ is zpd. This is a different statement than Theorem 3.10, since $M_n(F)^- \otimes B^-$ and $M_n(B)^-$ are different algebras (indeed we can identify the associative algebras $M_n(F) \otimes B$ and $M_n(B)$, but the algebras $M_n(F)^- \otimes B^-$ and $M_n(B)^-$ have different multiplications).

Finally, we show that Theorem 3.10 does not hold for all unital algebra B.

Example 3.13 For any $n \geq 2$, let B_{n^2+1} be the algebra from Example 3.9. We will prove that the matrix algebra $M_n(B_{n^2+1})$ is not zLpd.

As we know, there exists a bilinear functional φ on B_{n^2+1} such that

$$\sum_{t=1}^{n^2} [v_t, w_t] = 0 \implies \sum_{t=1}^{n^2} \varphi(v_t, w_t) = 0,$$

but there exist $u_t \in B_{n^2+1}, t = 1, \ldots, 2n^2 + 2$, satisfying

$$\sum_{t=1}^{n^2+1} [u_{2t-1}, u_{2t}] = 0 \quad \text{and} \quad \sum_{t=1}^{n^2+1} \varphi(u_{2t-1}, u_{2t}) \neq 0.$$

Define the bilinear functional Φ on $M_n(B_{n^2+1})$ by

$$\Phi(v, w) = \sum_{i=1}^{n} \sum_{j=1}^{n} \varphi(v_{ij}, w_{ji}),$$

where v_{ij} and w_{ij} are entries of the matrices v and w, respectively. Suppose that v and w commute. Then, in particular,

$$\sum_{j=1}^{n} v_{ij} w_{ji} = \sum_{j=1}^{n} w_{ij} v_{ji}$$

for all $i = 1, \ldots, n$, and hence

$$\sum_{i=1}^{n} \sum_{j=1}^{n} v_{ij} w_{ji} = \sum_{i=1}^{n} \sum_{j=1}^{n} w_{ij} v_{ji}.$$

Replacing the roles of i and j on the right-hand side, we see that this identity can be written as

$$\sum_{i=1}^{n} \sum_{j=1}^{n} [v_{ij}, w_{ji}] = 0.$$

Since this sum involves n^2 commutators, it follows that

$$\Phi(v, w) = \sum_{i=1}^{n} \sum_{j=1}^{n} \varphi(v_{ij}, w_{ji}) = 0.$$

Thus, Φ vanishes on pairs of commuting elements. However, it is not of the form $\Phi(x, y) = \tau([x, y])$ since the elements $v_t = u_{2t-1}e_{11}$ and $w_t = u_{2t}e_{11}, t = 1, \ldots, n^2+1,$ satisfy

$$\sum_{t=1}^{n^2+1} [v_t, w_t] = 0,$$

while

$$\sum_{t=1}^{n^2+1} \Phi(v_t, w_t) \neq 0.$$

This example in particular shows that algebras generated by idempotents are not always zLpd.

3.3   zJpd Algebras

The notion of a zero Jordan product determined algebra should by now be self-explanatory.

Definition 3.14 An associative algebra A is *zero Jordan product determined* if the Jordan algebra A^+ is zero product determined.

We will abbreviate "zero Jordan product determined" to *zJpd*.

We say that the algebra elements x and y *anticommute* if $x \circ y = 0$ (i.e., $xy = -yx$). Thus, an (associative) algebra A over a field F is zJpd if, for every bilinear functional $\varphi : A \times A \to F$ satisfying $\varphi(x, y) = 0$ whenever x and y anticommute, there exists a linear functional τ on A such that

$$\varphi(x, y) = \tau(x \circ y) \quad (x, y \in A). \tag{3.19}$$

If A is unital, then

$$\varphi(x, 1) = 2\tau(x) \quad (x \in A),$$

so (3.19) is equivalent to

$$\varphi(x, y) = \frac{1}{2}\varphi(x \circ y, 1) \quad (x, y \in A),$$

provided that the characteristic of F is not 2. This restriction on characteristic is quite common in the study of Jordan algebras. Under this assumption, a commutative algebra is zpd if and only if it is zJpd.

There are less obvious connections between zpd and zJpd algebras. We will show that unital algebras generated by idempotents, which are the main examples of zpd algebras, are also the main examples of zJpd algebras.

The following theorem was obtained in [24] (and partially in [85]).

Theorem 3.15 *Let A be an (associative) unital algebra over a field F of characteristic not 2. If A is generated by idempotents, then A is zJpd.*

Proof Let $\varphi : A \times A \to F$ be a bilinear functional with the property that $x \circ y = 0$ implies $\varphi(x, y) = 0$. Set

$$S = \{s \in A \mid \varphi(x, s) = \frac{1}{2}\varphi(x \circ s, 1) \text{ for all } x \in A\}. \tag{3.20}$$

Our goal is to show that $S = A$. Since S is a linear subspace of A and A is generated by idempotents, it is enough to show that S contains every element of the form $e_1 e_2 \ldots e_n$ where e_i are idempotents in A. We proceed by induction on n.

To establish the base case, take an idempotent $e = e_1 \in A$. We must prove that $e \in S$. Denote $1 - e$ by f. Writing

$$x = exe + fxe + exf + fxf,$$

we obtain

$$\varphi(x, e) = \varphi(exe, e) + \varphi(fxe, e) + \varphi(exf, e) + \varphi(fxf, e).$$

From

$$fxe \circ (2e - 1) = exf \circ (2e - 1) = fxf \circ e = 0$$

it follows that

$$\varphi(fxe, e) = \frac{1}{2}\varphi(fxe, 1),$$

$$\varphi(exf, e) = \frac{1}{2}\varphi(exf, 1),$$

$$\varphi(fxf, e) = 0.$$

Hence,

$$\varphi(x, e) = \varphi(exe, e) + \frac{1}{2}\varphi(fxe, 1) + \frac{1}{2}\varphi(exf, 1).$$

Further, $exe \circ f = 0$ and so $\varphi(exe, f) = 0$. Since $f = 1 - e$, we can write this as $\varphi(exe, e) = \varphi(exe, 1)$. Consequently,

$$\varphi(x, e) = \varphi(exe, 1) + \frac{1}{2}\varphi(fxe, 1) + \frac{1}{2}\varphi(exf, 1)$$

$$= \frac{1}{2}\varphi\big((exe + fxe) + (exe + exf), 1\big)$$

$$= \frac{1}{2}\varphi(x \circ e, 1).$$

That is, $e \in S$.

Assume now that S contains any product of n idempotents. Take idempotents $e_1, \ldots, e_n, e_{n+1}$. Our goal is to prove that S contains $e_1 \cdots e_n e_{n+1}$. Write

$$f_1 = 1 - e_1, \quad f_{n+1} = 1 - e_{n+1}, \quad t = e_2 \cdots e_n \quad (t = 1 \text{ if } n = 1).$$

Thus, we have to show that $e_1 t e_{n+1} \in S$. Take any $x \in A$. Using

$$x = e_{n+1} x e_1 + f_{n+1} x e_1 + e_{n+1} x f_1 + f_{n+1} x f_1,$$

we obtain

$$\varphi(x, e_1 t e_{n+1}) = \varphi(e_{n+1} x e_1, e_1 t e_{n+1})$$

$$+ \varphi(f_{n+1} x e_1, e_1 t e_{n+1})$$

$$+ \varphi(e_{n+1} x f_1, e_1 t e_{n+1})$$

$$+ \varphi(f_{n+1} x f_1, e_1 t e_{n+1})$$

$$= \varphi(e_{n+1} x e_1, t e_{n+1} - f_1 t + f_1 t f_{n+1})$$

$$+ \varphi(f_{n+1}xe_1, te_{n+1} - f_1te_{n+1})$$
$$+ \varphi(e_{n+1}xf_1, e_1t - e_1tf_{n+1})$$
$$+ \varphi(f_{n+1}xf_1, e_1te_{n+1}).$$

Since $e_{n+1}xe_1 \circ f_1tf_{n+1} = 0$, we have

$$\varphi(e_{n+1}xe_1, f_1tf_{n+1}) = 0.$$

Similarly,

$$\varphi(f_{n+1}xe_1, f_1te_{n+1}) = 0,$$

$$\varphi(e_{n+1}xf_1, e_1tf_{n+1}) = 0,$$

$$\varphi(f_{n+1}xf_1, e_1te_{n+1}) = 0.$$

Hence,

$$\varphi(x, e_1te_{n+1}) = \varphi(e_{n+1}xe_1, te_{n+1} - f_1t)$$
$$+ \varphi(f_{n+1}xe_1, te_{n+1})$$
$$+ \varphi(e_{n+1}xf_1, e_1t).$$

Since each of the elements te_{n+1}, f_1t, te_{n+1}, and e_1t is the product of n idempotents, the induction assumption implies that

$$\varphi(e_{n+1}xe_1, te_{n+1}) = \frac{1}{2}\varphi(e_{n+1}xe_1te_{n+1} + te_{n+1}xe_1, 1),$$

$$\varphi(e_{n+1}xe_1, f_1t) = \frac{1}{2}\varphi(f_1te_{n+1}xe_1, 1),$$

$$\varphi(f_{n+1}xe_1, te_{n+1}) = \frac{1}{2}\varphi(f_{n+1}xe_1te_{n+1}, 1),$$

$$\varphi(e_{n+1}xf_1, e_1t) = \frac{1}{2}\varphi(e_1te_{n+1}xf_1, 1).$$

Therefore,

$$\varphi(x, e_1 t e_{n+1})$$

$$= \frac{1}{2}\varphi(e_{n+1}xe_1 t e_{n+1} + t e_{n+1}xe_1 - f_1 t e_{n+1}xe_1 + f_{n+1}xe_1 t e_{n+1} + e_1 t e_{n+1}xf_1, 1)$$

$$= \frac{1}{2}\varphi(x \circ e_1 t e_{n+1}, 1).$$

This proves that $e_1 t e_{n+1} \in S$. $\square$

Remark 3.16 From the proof of Theorem 3.15 we see that the following slightly more general result holds: If A is a unital algebra and $\varphi : A \times A \to F$ is a bilinear functional with the property that $\varphi(x, y) = 0$ whenever $x \circ y = 0$, then the set S defined in (3.20) contains the subalgebra generated by all idempotents in A.

Theorem 3.15 yields the following analogue of Corollary 2.17.

Corollary 3.17 *The following algebras over a field of characteristic not 2 are zJpd:*

(a) *The algebra $M_n(B)$, where $n \geq 2$ and B is any unital algebra.*
(b) *The algebra $\mathrm{Tri}(A_1, M, A_2)$, where A_1 and A_2 are unital algebras generated by idempotents and M is a unital (A_1, A_2)-bimodule.*

The assumption that the characteristic of F is not 2 is really necessary here. Example 3.13 shows that the matrix algebras $M_n(B)$ are not always zpd with respect to the Lie product. However, if the characteristic is 2, then the Lie product coincides with the Jordan product.

Let us now look for some non-examples. In light of Corollary 1.8, every zJpd algebra of dimension greater than 1 contains pairs of nonzero anticommuting elements. The easiest way to find non-examples of zJpd algebras is therefore searching for algebras in which $x \circ y = 0$ implies $x = 0$ or $y = 0$. Among commutative algebras over fields of characteristic not 2, these are exactly the algebras without zero-divisors. The noncommutative context is of course more intriguing. In the next example, we show that one of most prominent noncommutative algebras has this property.

Example 3.18 The (first) Weyl algebra A is the complex algebra defined by the generators x and y and the relation $xy - yx = 1$. It is an elementary exercise to show that every element in A can be uniquely written in the form $\sum_{i=0}^{m} f_i(x)y^i$ for some polynomials f_i, and that, for any polynomials f_i and g_j,

$$\left(\sum_{i=0}^{m} f_i(x)y^i\right)\left(\sum_{j=0}^{n} g_j(x)y^j\right) = f_m(x)g_n(x)y^{m+n} + \sum_{\ell=0}^{m+n-1} h_\ell(x)y^\ell$$

for some polynomials h_ℓ (see [47, Example 2.28]). Therefore,

$$\left(\sum_{i=0}^{m} f_i(x)y^i\right) \circ \left(\sum_{j=0}^{n} g_j(x)y^j\right) = 2f_m(x)g_n(x)y^{m+n} + \sum_{\ell=0}^{m+n-1} k_\ell(x)y^\ell$$

for some polynomials k_ℓ. This shows that A has no nonzero anticommuting elements. In particular, it is not zJpd.

In order to search for examples of zJpd algebras that are not generated by idempotents, it seems natural to look at algebras that have many pairs of anticommuting elements. One such algebra is the algebra of real *quaternions* $\mathbb{H}$ whose standard basis consists of 1 and three elements that anticommute with each other.

Example 3.19 One way of looking at $\mathbb{H}$ is that this is the real vector space $\mathbb{R} \times \mathbb{R}^3$ endowed with multiplication given by

$$(\alpha, \mathbf{x})(\beta, \mathbf{y}) = (\alpha\beta - \mathbf{x} \cdot \mathbf{y}, \alpha\mathbf{y} + \beta\mathbf{x} + \mathbf{x} \times \mathbf{y}),$$

where $\mathbf{x} \cdot \mathbf{y}$ and $\mathbf{x} \times \mathbf{y}$ are the dot product and the vector product of $\mathbf{x}$ and $\mathbf{y}$, respectively. If

$$(\alpha, \mathbf{x}) \circ (\beta, \mathbf{y}) = 0,$$

then $\alpha\beta - \mathbf{x} \cdot \mathbf{y} = 0$ and $\alpha\mathbf{y} + \beta\mathbf{x} = 0$, and hence

$$\alpha^2\beta^2 = \beta\mathbf{x} \cdot \alpha\mathbf{y} = -\beta\mathbf{x} \cdot \beta\mathbf{x}.$$

Since $\alpha^2\beta^2 \geq 0$ and $-\beta\mathbf{x} \cdot \beta\mathbf{x} \leq 0$, this is possible only if $\alpha = 0$ or $\beta = 0$. Hence, the bilinear functional $\varphi \colon \mathbb{H} \times \mathbb{H} \to \mathbb{R}$ defined by

$$\varphi\big((\alpha, \mathbf{x}), (\beta, \mathbf{y})\big) = \alpha\beta$$

vanishes on each pair of anticommuting elements. However, since

$$(1, 0) \circ (1, 0) + (0, \mathbf{i}) \circ (0, \mathbf{i}) = 0$$

but

$$\varphi\big((1,0),(1,0)\big) + \varphi\big((0,\mathbf{i}),(0,\mathbf{i})\big) = 1 \cdot 1 + 0 \cdot 0 = 1,$$

φ does not satisfy condition (ii) of Proposition 1.3. Therefore, $\mathbb{H}$ is not zJpd.

Another well-known example of an algebra with plenty of pairs of anticommuting elements is the *Grassmann (exterior) algebra*.

Example 3.20 Let F be a field of characteristic not 2. The Grassmann algebra G_n over F with n generators $x_1, \ldots, x_n$ is the 2^n-dimensional algebra whose standard basis consists of unity 1 and elements of the form $x_{i_1} \cdots x_{i_k}$, with $1 \le k \le n$ and $i_1 < \cdots < i_k$, whose multiplication is determined by $x_i^2 = 0$ and $x_i \circ x_j = 0$ for all i and j. Suppose that two elements in G_n,

$$\alpha_0 + \sum \alpha_{i_1,\ldots,i_k} x_{i_1} \cdots x_{i_k} \text{ and } \beta_0 + \sum \beta_{i_1,\ldots,i_k} x_{i_1} \cdots x_{i_k},$$

anticommute. Observe that then $\alpha_0 \beta_0 = 0$ and $\alpha_0 \beta_1 + \alpha_1 \beta_0 = 0$. The bilinear map given by

$$\varphi\left(\alpha_0 + \sum \alpha_{i_1,\ldots,i_k} x_{i_1} \cdots x_{i_k}, \ \beta_0 + \sum \beta_{i_1,\ldots,i_k} x_{i_1} \cdots x_{i_k}\right) = \alpha_0 \beta_1$$

therefore vanishes on pairs of anticommuting elements. However, since $\varphi(1, x_1) \ne \varphi(x_1, 1)$, φ is not of the form $\varphi(x, y) = \tau(x \circ y)$. Therefore, G_n is not zJpd. (The same proof shows that Grassmann algebras in infinitely many generators are not zJpd.)

These two examples show that the abundance of pairs of anticommuting elements in A does not yet imply that A is zJpd. As a matter of fact, we are facing the same problem as with zpd algebras: we do not know of any unital zJpd algebra (over a field of characteristic not 2) that is not generated by idempotents. Generally speaking, our understanding of zJpd algebras is, at present, rather limited.

The rest of the section is devoted to a variation of the notion of a zJpd algebra. Note first that since Jordan algebras are commutative, the notion of a symmetrically zpd Jordan algebra makes sense. Again, we will consider only Jordan algebras of the form A^+ where A is an associative algebra.

Definition 3.21 An associative algebra A is *symmetrically zero Jordan product determined* (*symmetrically zJpd* for short) if the Jordan algebra A^+ is symmetrically zero product determined.

Thus, A is symmetrically zJpd if, for every symmetric bilinear functional φ on A satisfying $\varphi(x, y) = 0$ whenever x and y anticommute, there exists a linear functional τ on A such that $\varphi(x, y) = \tau(x \circ y)$ for all $x, y \in A$. If A is zJpd, then this condition is fulfilled even without assuming in advance that φ is symmetric, so A is also symmetrically zJpd. To show that the converse is not true, we point out that (unlike in Example 3.19) the bilinear functional φ in Example 3.20 is not symmetric. This raises the question whether the Grassmann algebra G_n, which we know is not zJpd, is symmetrically zJpd.

Example 3.22 The Jordan algebra G_2^+ is actually associative. More precisely, the Jordan product of any two elements in span $\{x_1, x_2, x_1 x_2\}$ is 0 and G_2^+ is isomorphic to the algebra $A^\#$ from Example 2.23, with A being a 3-dimensional algebra with trivial multiplication. Therefore, G_2^+ is symmetrically zpd. So, G_2 is an example of a symmetrically zJpd algebra that is not zJpd.

On the other hand, the algebras G_n with $n \geq 3$ are not symmetrically zJpd. To show this, consider the symmetric bilinear functional defined by

$$\varphi\left(\alpha_0 + \sum \alpha_{i_1,\dots,i_k} x_{i_1} \cdots x_{i_k}, \; \beta_0 + \sum \beta_{i_1,\dots,i_k} x_{i_1} \cdots x_{i_k}\right) = \alpha_0 \beta_{1,\dots,n} + \alpha_{1,\dots,n} \beta_0.$$

Observe that

$$\left(\alpha_0 + \sum \alpha_{i_1,\dots,i_k} x_{i_1} \cdots x_{i_k}\right) \circ \left(\beta_0 + \sum \beta_{i_1,\dots,i_k} x_{i_1} \cdots x_{i_k}\right) = 0$$

implies that either one of the factors is 0 or $\alpha_0 = \beta_0 = 0$. Therefore, φ vanishes on pairs of anticommuting elements. However,

$$x_1 x_2 \circ x_3 \cdots x_n = 1 \circ x_1 \cdots x_n$$

while

$$\varphi(x_1 x_2, x_3 \cdots x_n) \neq \varphi(1, x_1 \cdots x_n),$$

so φ is not of the form $\varphi(x, y) = \tau(x \circ y)$.

We do not know whether there is some direct connection between the classes of zpd and zJpd algebras. What can be shown, however, is that, under mild assumptions, zpd algebras are also symmetrically zJpd. The next theorem tells us more than just that.

Theorem 3.23 *Let A be an (associative) unital zpd algebra over a field F. If a bilinear functional $\varphi : A \times A \to F$ satisfies $\varphi(x, y) = 0$ whenever $xy = yx = 0$, then*

$$\varphi(xy, zw) + \varphi(wx, yz) = \varphi(x, yzw) + \varphi(wxy, z) \quad (x, y, z, w \in A). \tag{3.21}$$

Moreover, if φ is symmetric, then

$$2\varphi(x, y) = \varphi(x \circ y, 1) \quad (x, y \in A). \tag{3.22}$$

Proof Take any $z, w \in A$ such that $zw = 0$ and define the bilinear functional φ' on A by

$$\varphi'(x, y) = \varphi(wx, yz).$$

Observe that $xy = 0$ implies $wx \cdot yz = yz \cdot wx = 0$ and hence $\varphi'(x, y) = 0$. Since A is zpd, it follows that

$$\varphi'(x, y) = \varphi'(xy, 1) \quad (x, y \in A).$$

That is to say, $zw = 0$ implies $\varphi(wx, yz) - \varphi(wxy, z) = 0$ for all $x, y \in A$. The assumption that A is zpd can therefore be applied to the map

$$\varphi''(z, w) = \varphi(wx, yz) - \varphi(wxy, z),$$

where x and y an arbitrary but fixed elements in A. Hence we obtain

$$\varphi''(z, w) = \varphi''(zw, 1) \quad (z, w \in A),$$

which is nothing but (3.21). Setting $x = z = 1$ we get

$$\varphi(y, w) + \varphi(w, y) = \varphi(1, yw) + \varphi(wy, 1) \quad (y, w \in A).$$

If φ is symmetric, this gives (3.22).   □

Remark 3.24 In light of Proposition 1.3 (iv), it is clear from the proof that Theorem 3.23 holds for bilinear maps φ from $A \times A$ to any vector space X, not only for bilinear functionals φ.

Corollary 3.25 *Let A be an (associative) unital algebra over a field of characteristic not 2. If A is zpd, then A is also symmetrically zJpd.*

Proof One only has to use the obvious fact that $xy = yx = 0$ implies $x \circ y = 0$.   □

The converse of Corollary 3.25 does not hold.

Example 3.26 The Grassmann algebra G_2 is symmetrically zJpd (Example 3.22), but is not zpd since it is not generated by idempotents (see Theorem 2.27).

If the characteristic of F is not 2, Theorem 3.23 obviously gives the best possible conclusion for the case where φ is symmetric. What about if φ is skew-symmetric? The expected conclusion would be that there exists a linear functional τ on A such that $\varphi(x, y) = \tau([x, y])$ for all $x, y \in A$. This turns out to be true if $A = M_n(D)$, $n \geq 2$, with D a finite-dimensional central division algebra [115]. However, as the next two examples show, it is not true in general (which also shows that the assumption that the characteristic of F is not 2 is necessary in the symmetric case). The first example is just a reinterpretation of Example 3.13.

Example 3.27 The bilinear functional Φ from Example 3.13 vanishes on pairs of commuting elements, so in particular $\Phi(x, y) = 0$ whenever $xy = yx = 0$. Moreover, $\Phi(x, x) = 0$ which implies that Φ is skew-symmetric. However, Φ is not of the form $\Phi(x, y) = \tau([x, y])$, although $M_n(B_{n^2+1})$ is a unital zpd algebra (by Corollary 2.17).

The next example, which was discovered by Žan Bajuk, gives a deeper insight.

Example 3.28 Let F be a field of characteristic 0, let $B = F \oplus Fu$ be the 2-dimensional unital algebra over F determined by $u^2 = 0$, and let $A = M_n(B)$, $n \geq 2$. Writing any $x \in A$ as $x = x_0 + ux_1$ with $x_0, x_1 \in M_n(F)$, we have

$$xy = (x_0 + ux_1)(y_0 + uy_1) = x_0 y_0 + u(x_0 y_1 + x_1 y_0).$$

Define $\varphi : A \times A \to F$ by

$$\varphi(x, y) = \mathrm{tr}(x_0 y_1 - x_1 y_0),$$

where $\mathrm{tr}(z)$ stands for the trace of the matrix $z \in M_n(F)$. Obviously, φ is a skew-symmetric bilinear functional. Suppose $x, y \in A$ satisfy $xy = yx = 0$. Then

$$x_0 y_0 = 0, \quad y_0 x_0 = 0, \quad x_0 y_1 + x_1 y_0 = 0, \quad \text{and} \quad y_0 x_1 + y_1 x_0 = 0.$$

Observe that the first and the last identity yield

$$(x_0 y_1)^2 = x_0(y_1 x_0)y_1 = -(x_0 y_0)x_1 y_1 = 0.$$

Thus, $x_0 y_1 = 0$ is a nilpotent matrix and so its trace is 0. As $x_0 y_1 = -x_1 y_0$ by the third identity, we see that $\varphi(x, y) = 0$. However, not only that φ is not of the form $\varphi(x, y) = \tau([x, y])$, but does not even vanish on pairs of commuting elements. In fact,

$$\varphi(1, u1) = \mathrm{tr}(1) = n \neq 0,$$

where 1 stands for the identity matrix.

Recall from Proposition 2.7 that the condition that a unital algebra A is zpd is equivalent to the condition that any bilinear functional $\varphi : A \times A \to F$ with the property

$$xy = 0 \implies \varphi(x, y) = 0 \quad (x, y \in A)$$

satisfies

$$\varphi(x, 1) = \varphi(1, x) \quad (x \in A). \tag{3.23}$$

Example 3.28 in particular shows that, in a unital zpd algebra A, a bilinear functional $\varphi : A \times A \to F$ which satisfies

$$xy = yx = 0 \implies \varphi(x, y) = 0 \quad (x, y \in A) \tag{3.24}$$

does not necessarily satisfy (3.23) (which reads as $\varphi(x, 1) = 0$ if φ is skew-symmetric and F has characteristic different from 2). This is slightly disappointing in view of possible applications.

We will continue the discussion on condition (3.24) in Sect. 6.1. Let us mention at the end another variation of the notion of a zJpd algebra. An algebra A is said to be *square-zero determined* if every symmetric bilinear functional $\varphi : A \times A \to F$ satisfying $\varphi(x, x) = 0$ whenever $x^2 = 0$ is of the form $\varphi(x, y) = \tau(x \circ y)$ with τ a linear functional. This concept was studied in [127, 166].

Part II
Analytic Theory

Zero Product Determined Nonassociative Banach Algebras

4

Part II studies the zpd property in Banach algebras. We will soon see that the algebraic theory, presented in Part I, has rather limited reach in this context. By requiring that the functionals from the algebraic definition of a zpd algebra are continuous, we obtain the "right" definition of a zpd Banach algebra. The main goal of Part II is to show that the class of such Banach algebras is large.

The organization of Part II is similar to that of Part I. In this introductory chapter, we work in the most general setting in which we can define the basic concept.

4.1 Characters and the Limitations of the Algebraic Approach

First, we fix some notation and terminology. Throughout Part II, all algebras and all vector spaces are taken over the complex field $\mathbb{C}$. By X' we denote the dual space of the normed space X. For $S \subseteq X$, span S and $\overline{\text{span}}\, S$ denote the linear span and the closed linear span of S, respectively.

As in Part I, we use the term nonassociative algebra in order to emphasize that the associativity of the product is not being assumed; we do not mean that the associativity fails to hold, but only that the associativity does not need to hold. A *nonassociative Banach algebra* is a nonassociative complex algebra A whose underlying vector space is a Banach space with respect to a norm $\| \cdot \|$ which is related to the algebraic structure of A by the requirement that

$$\|xy\| \leq \|x\|\|y\| \quad (x, y \in A).$$

In the case where A has a unity, 1, we also require that $\|1\| = 1$.

© The Author(s), under exclusive license to Springer Nature Switzerland AG 2021
M. Brešar, *Zero Product Determined Algebras*, Frontiers in Mathematics,
https://doi.org/10.1007/978-3-030-80242-4_4

The most important classes of nonassociative Banach algebras are Banach algebras, Lie–Banach algebras, and Jordan–Banach algebras. As usual, the term *Banach algebra* is reserved for an associative Banach algebra. By a *Lie–Banach algebra* we mean a nonassociative Banach algebra whose underlying algebra is a Lie algebra. Similarly, by a *Jordan–Banach algebra* we mean a nonassociative Banach algebra whose underlying algebra is a Jordan algebra. For every Banach algebra A, the Lie algebra A^- becomes a Lie–Banach algebra and the Jordan algebra A^+ becomes a Jordan–Banach algebra, both of them with respect to the norm $2\| \cdot \|$.

Let A be a nonassociative Banach algebra. A *character* on A is a nonzero multiplicative linear functional on A. The *character space* Ω_A of A is the set of characters on A. Each $\omega \in \Omega_A$ is continuous with $\|\omega\| \leq 1$. Indeed, assume toward a contradiction that there exists an $x \in A$ with $\|x\| < 1$ and $\omega(x) = 1$. Let L_x be the left multiplication operator (given by $L_x(u) = xu$). Then $\|L_x\| \leq \|x\| < 1$, so the series

$$\sum_{n=0}^{\infty} L_x^n(x)$$

converges in A, say to y. Clearly $y - xy = x$, and hence

$$\omega(y) - \omega(x)\omega(y) = \omega(x),$$

which is impossible since $\omega(x) = 1$. Thus, $|\omega(x)| \leq 1$ whenever $x \in A$ is such that $\|x\| = 1$.

Definition 4.1 A nonassociative algebra A is *third power-associative* if $x^2x = xx^2$ for every $x \in A$.

Note that associative, Lie, and Jordan algebras are third power-associative.

Theorem 4.2 *Let A be a third power-associative (nonassociative) Banach algebra. If A has infinitely many characters, then A is not a zpd algebra.*

Proof Let $\{\omega_n \mid n \in \mathbb{N}\}$ be a family of distinct characters on A. For $m, n \in \mathbb{N}, m \neq n$, set

$$S_{m,n} = \{a \in A \mid 0 \neq \omega_m(a) \neq \omega_n(a) \neq 0\}.$$

Observe that $A \setminus S_{m,n}$ is the union of the following three sets:

$$A_1 = \{a \in A \mid (\omega_m - \omega_n)(a) = 0\},$$

$$A_2 = \{a \in A \mid \omega_m(a) = 0\},$$

$$A_3 = \{a \in A \mid \omega_n(a) = 0\}.$$

As A_1, A_2, A_3 are closed hyperplanes, $S_{m,n}$ is an open dense subset of A. By the Baire category theorem,

$$S = \bigcap_{m \neq n} S_{m,n}$$

is dense in A.

Take any $a \in S$. Our goal is to prove that

$$a^2 \otimes a - a \otimes a^2 \notin \text{span}\{x \otimes y \mid x, y \in A, \; xy = 0\}. \tag{4.1}$$

Suppose this is not true. Then

$$a^2 \otimes a - a \otimes a^2 = \sum_{j=1}^{k} x_j \otimes y_j$$

for some $x_1, \dots, x_k, y_1, \dots y_k \in A$ satisfying $x_j y_j = 0$, $1 \leq j \leq k$. We thus have

$$\omega_m(a)\omega_n(a)\big(\omega_m(a) - \omega_n(a)\big)$$

$$= (\omega_m \otimes \omega_n)(a^2 \otimes a - a \otimes a^2)$$

$$= \sum_{j=1}^{k} (\omega_m \otimes \omega_n)(x_j \otimes y_j) \tag{4.2}$$

$$= \sum_{j=1}^{k} \omega_m(x_j)\omega_n(y_j)$$

for all $m, n \in \mathbb{N}$.

Since $x_j y_j = 0$, it follows that

$$\{n \in \mathbb{N} \mid \omega_n(x_j) = 0\} \cup \{n \in \mathbb{N} \mid \omega_n(y_j) = 0\} = \mathbb{N}.$$

Therefore, we can successively choose $z_j \in \{x_j, y_j\}$ such that the set

$$\{n \in \mathbb{N} \mid \omega_n(z_1) = \cdots = \omega_n(z_j) = 0\}$$

is infinite. This process stops when k is reached, and we then have that the set

$$\{n \in \mathbb{N} \mid \omega_n(z_1) = \cdots = \omega_n(z_k) = 0\}$$

is infinite. In particular, by taking $m \neq n$ from this set we obtain

$$\omega_m(x_1)\omega_n(y_1) = \cdots = \omega_m(x_k)\omega_n(y_k) = 0.$$

From (4.2) we now deduce that

$$\omega_m(a)\omega_n(a)\big(\omega_m(a) - \omega_n(a)\big) = 0,$$

which contradicts the assumption that $a \in S$.

Since, by the assumption of the theorem, $a^2a - aa^2 = 0$, we see from (4.1) that A does not satisfy condition (iii) of Proposition 1.3. Therefore, A is not zpd.   □

Remark 4.3 The direct sum of infinitely many copies of $\mathbb{C}$ is a zpd algebra by Theorem 1.16 and has infinitely many characters (given by $\omega_n\big((x_i)\big) = x_n$). However, this normed algebra is not a Banach algebra.

Remark 4.4 If a unital Banach algebra A is generated by idempotents (as an algebra, not as a Banach algebra), then it is a zpd algebra by Thcorem 2.15. Theorem 4.2 therefore implies that such an algebra does not have infinitely many characters. (Of course, it can have finitely many characters, just think of the direct sum of finitely many copies of $\mathbb{C}$.)

Recall that a commutative Banach algebra A is said to be *semisimple* if

$$\bigcap_{\omega \in \Omega_A} \ker \omega = \{0\}.$$

The following corollary follows immediately from Theorem 4.2 together with the standard Gelfand theory of commutative Banach algebras.

Corollary 4.5 *Let A be an infinite-dimensional, commutative, semisimple Banach algebra. Then A is not a zpd algebra.*

It should be pointed out that A is not zpd only as an algebra, which does not necessarily mean that it is not zpd as a Banach algebra in the sense of Definition 4.6 below (see also Remark 4.8). What Corollary 4.5 shows is that the algebraic theory of zpd algebras is inapplicable to a particularly important, basic class of Banach algebras. One is therefore forced to seek for a different approach.

4.2   The Definition of a zpd Nonassociative Banach Algebra

We just saw that the algebraic definition of a zpd algebra is not suitable for the Banach algebra setting. Adding the requirement that the functionals from this definition are continuous makes a big difference, as we shall see.

Definition 4.6 A nonassociative Banach algebra A is said to be *zero product determined*, or *zpd* for short, if, for every continuous bilinear functional $\varphi \colon A \times A \to \mathbb{C}$ satisfying $\varphi(x, y) = 0$ whenever $x, y \in A$ are such that $xy = 0$, there exists a continuous linear functional τ on A such that $\varphi(x, y) = \tau(xy)$ for all $x, y \in A$.

To prevent misunderstanding, we add two remarks to this definition.

Remark 4.7 Recall once again that "nonassociative" means "not necessarily associative" (and not "not associative"). Definition 4.6 therefore also covers the ordinary, associative Banach algebras, in which we are in fact primarily interested.

Remark 4.8 Let us emphasize that, for a nonassociative Banach algebra A, "A is a zpd algebra" and "A is a zpd Banach algebra" are different statements. In particular, we will see later that many of the algebras from Corollary 4.5 are zpd Banach algebras. Throughout Part II, the statement "A is zpd" should be understood as that A is a zpd (nonassociative) Banach algebra.

We continue to assume that A is a nonassociative Banach algebra. The product map $(x, y) \mapsto xy$ is a continuous bilinear map from $A \times A$ into A, so there is a continuous linear map π from $A \widehat{\otimes} A$, the projective tensor product of two copies of A, to A satisfying

$$\pi(x \otimes y) = xy$$

for all $x, y \in A$. Note that π induces a continuous linear bijection

$$\widetilde{\pi} \colon A \widehat{\otimes} A / \ker \pi \to \pi(A \widehat{\otimes} A) \subseteq \overline{A^2}$$

which is an isomorphism onto $\overline{A^2}$ if and only if it satisfies any of the following equivalent properties:

1. there exists a $C > 0$ such that $\|u + \ker \pi\| \leq C \|\pi(u)\|$ for each $u \in A \widehat{\otimes} A$;
2. $\pi(A \widehat{\otimes} A)$ is closed in A;
3. $\pi(A \widehat{\otimes} A) = \overline{A^2}$.

We continue with a "Banach version" of Proposition 1.3.

Proposition 4.9 *Let A be a nonassociative Banach algebra. The following conditions are equivalent:*

(i) *A is zpd.*

(ii) *If a continuous bilinear functional $\varphi \colon A \times A \to \mathbb{C}$ has the property that $\varphi(x, y) = 0$ whenever $x, y \in A$ are such that $xy = 0$, then there exists a $C > 0$ such that*

$$|\sum_{k=1}^{n} \varphi(x_k, y_k)| \le C \| \sum_{k=1}^{n} x_k y_k \| \quad (x_1, \ldots, x_n, y_1, \ldots, y_n \in A).$$

(iii) $\ker \pi = \overline{\text{span}} \{x \otimes y \in A\widehat{\otimes}A \mid xy = 0\}$ *and $\pi(A\widehat{\otimes}A)$ is closed in A.*

(iv) *For every normed space X and every continuous bilinear map $\varphi \colon A \times A \to X$ satisfying $\varphi(x, y) = 0$ whenever $x, y \in A$ are such that $xy = 0$, there exists a continuous linear map $T \colon A^2 \to X$ such that $\varphi(x, y) = T(xy)$ for all $x, y \in A$.*

Proof (i) $\implies$ (iv). Let X be a normed space and let $\varphi \colon A \times A \to X$ be a continuous bilinear map as in (iv). For each $\xi \in X'$, $\xi \circ \varphi$ is a continuous bilinear functional on A that satisfies the condition to which (i) is applicable. Therefore, there exists a $\tau(\xi) \in A'$ such that

$$(\xi \circ \varphi)(x, y) = \tau(\xi)(xy)$$

for all $x, y \in A$. Of course, $\tau(\xi)$ is the unique functional in $(A^2)'$ with this property. Thus, the map $\tau \colon X' \to (A^2)'$ is well-defined and linear. We next show that τ is continuous. Let (ξ_n) be a sequence in X' with $\lim \xi_n = 0$ and $\lim \tau(\xi_n) = \xi$ for some $\xi \in (A^2)'$. For any $x, y \in A$, we have

$$0 = \lim \xi_n(\varphi(x, y)) = \lim \tau(\xi_n)(xy) = \xi(xy).$$

We thus have $\xi = 0$, and the closed graph theorem yields the continuity of τ.

Next, for all $x_i, y_i \in A$, $i = 1, \ldots, n$, and $\xi \in X'$, we have

$$\xi\left(\sum_{k=1}^{n} \varphi(x_k, y_k)\right) = \sum_{k=1}^{n} \xi\big(\varphi(x_k, y_k)\big)$$

$$= \sum_{k=1}^{n} \tau(\xi)(x_k y_k) \tag{4.3}$$

$$= \tau(\xi)\left(\sum_{k=1}^{n} x_k y_k\right).$$

Hence, if $\sum_{k=1}^{n} x_k y_k = 0$, then

$$\xi\Big(\sum_{k=1}^{n} \varphi(x_k, y_k)\Big) = 0$$

for each $\xi \in X'$, and therefore $\sum_{k=1}^{n} \varphi(x_k, y_k) = 0$. We can thus define a linear map $T \colon A^2 \to X$ by

$$T\Big(\sum_{k=1}^{n} x_k y_k\Big) = \sum_{k=1}^{n} \varphi(x_k, y_k)$$

for all $x_i, y_i \in A$, $i = 1, \ldots, n$. Of course, $\varphi(x, y) = T(xy)$ for all $x, y \in A$.

Our next concern is the continuity of T. Given $x_i, y_i \in A$, $i = 1, \ldots, n$, there exists a $\xi \in X'$ with $\|\xi\| = 1$ such that

$$\xi\Big(\sum_{k=1}^{n} \varphi(x_k, y_k)\Big) = \Big\|\sum_{k=1}^{n} \varphi(x_k, y_k)\Big\|.$$

On account of (4.3), we have

$$\Big\|T\Big(\sum_{k=1}^{n} x_k y_k\Big)\Big\| = \Big\|\sum_{k=1}^{n} \varphi(x_k, y_k)\Big\|$$

$$= \xi\Big(\sum_{k=1}^{n} \varphi(x_k, y_k)\Big)$$

$$= \Big|\tau(\xi)\Big(\sum_{k=1}^{n} x_k y_k\Big)\Big|$$

$$\leq \|\tau(\xi)\|\Big\|\sum_{k=1}^{n} x_k y_k\Big\|$$

$$\leq \|\tau\|\Big\|\sum_{k=1}^{n} x_k y_k\Big\|,$$

which proves the continuity of T. Thus, (iv) holds.

(iv) $\Longrightarrow$ (iii). Set

$$W = \overline{\mathrm{span}}\,\{x \otimes y \in A\widehat{\otimes}A \mid xy = 0\}.$$

If $x, y \in A$ are such that $xy = 0$, then $\pi(x \otimes y) = 0$, and therefore $W \subseteq \ker \pi$. Assume that $W \neq \ker \pi$. Then there exists a $\xi \in (A \widehat{\otimes} A)'$ such that $\xi(W) = \{0\}$ and $\xi(\ker \pi) \neq \{0\}$. Define a continuous bilinear functional $\varphi \colon A \times A \to \mathbb{C}$ by

$$\varphi(x, y) = \xi(x \otimes y).$$

If $x, y \in A$ are such that $xy = 0$, then $x \otimes y \in W$ and so $\varphi(x, y) = 0$. By hypothesis, there exists a continuous linear functional $\tau \colon A^2 \to \mathbb{C}$ such that $\varphi(x, y) = \tau(xy)$ for all $x, y \in A$. This functional extends to a continuous linear functional on A, which we also denote by τ. We thus have

$$\xi(x \otimes y) = \varphi(x, y) = \tau(xy) = (\tau \circ \pi)(x \otimes y) \quad (x, y \in A).$$

Since both ξ and $\tau \circ \pi$ are continuous, it follows that

$$\xi(u) = (\tau \circ \pi)(u) \quad (u \in A \widehat{\otimes} A).$$

This clearly implies that $\xi(\ker \pi) = \{0\}$, which is a contradiction.

We now consider the continuous bilinear map $\psi \colon A \times A \to A \widehat{\otimes} A / \ker \pi$ defined by

$$\psi(x, y) = x \otimes y + \ker \pi.$$

It is clear that $\psi(x, y) = 0$ whenever $x, y \in A$ are such that $xy = 0$. Therefore, by hypothesis, there exists a continuous linear map $T \colon A^2 \to A \widehat{\otimes} A / \ker \pi$ such that

$$\psi(x, y) = T(xy) \quad (x, y \in A).$$

Since the quotient space $A \widehat{\otimes} A / \ker \pi$ is complete, the map T extends uniquely to a continuous linear map from $\overline{A^2}$ to $A \widehat{\otimes} A / \ker \pi$. We continue to write T for this extension. We thus have

$$x \otimes y + \ker \pi = T(\pi(x \otimes y)) \quad (x, y \in A).$$

As T is continuous, we have $u + \ker \pi = T(\pi(u))$ for each $u \in A \widehat{\otimes} A$, which gives

$$\|u + \ker \pi\| \leq \|T\| \|\pi(u)\|.$$

As pointed out before the statement of the proposition, this means that $\pi(A \widehat{\otimes} A)$ is closed in A.

(iii) $\implies$ (ii). Take a continuous bilinear functional $\varphi \colon A \times A \to \mathbb{C}$ satisfying $\varphi(x, y) = 0$ whenever $xy = 0$. Let $\widehat{\varphi}$ be a continuous linear functional on $A \widehat{\otimes} A$ such that

$$\widehat{\varphi}(x \otimes y) = \varphi(x, y) \quad (x, y \in A).$$

Since $\varphi(x, y) = 0$ whenever $x, y \in A$ are such that $xy = 0$, it follows that $\widehat{\varphi}$ vanishes on $\overline{\operatorname{span}}\{x \otimes y \in A \widehat{\otimes} A \mid xy = 0\}$, which, by hypothesis, coincides with $\ker \pi$. Consequently, $\widehat{\varphi}$ induces a continuous linear functional $\widetilde{\varphi}$ on the quotient $A \widehat{\otimes} A / \ker \pi$. Since $\pi(A \widehat{\otimes} A)$ is closed in A, $\widetilde{\pi}$ is a linear isomorphism from $A \widehat{\otimes} A / \ker \pi$ onto $\overline{A^2}$. Then, for all $x_1, \ldots, x_n, y_1, \ldots, y_n \in A$, we have

$$\left| \sum_{k=1}^{n} \varphi(x_k, y_k) \right| = \left| \widehat{\varphi} \left(\sum_{k=1}^{n} x_k \otimes y_k \right) \right|$$

$$= \left| \widetilde{\varphi} \left(\sum_{k=1}^{n} x_k \otimes y_k + \ker \pi \right) \right|$$

$$= \left| (\widetilde{\varphi} \circ \widetilde{\pi}^{-1}) \left(\sum_{k=1}^{n} x_k \otimes y_k \right) \right|$$

$$\leq \left\| \widetilde{\varphi} \circ \widetilde{\pi}^{-1} \right\| \left\| \sum_{k=1}^{n} x_k \otimes y_k \right\|,$$

as desired.

(ii) $\implies$ (i) Let $\varphi \colon A \times A \to \mathbb{C}$ be a continuous bilinear functional such that $\varphi(x, y) = 0$ whenever $x, y \in A$ are such that $xy = 0$. Let $C > 0$ be the constant from (ii), so that

$$\left| \sum_{k=1}^{n} \varphi(x_k, y_k) \right| \leq C \left\| \sum_{k=1}^{n} x_k y_k \right\|$$

for all $x_1, \ldots, x_n, y_1, \ldots, y_n \in A$. This property allows us to define a continuous linear functional $\tau \colon A^2 \to \mathbb{C}$ by

$$\tau \left(\sum_{k=1}^{n} x_k y_k \right) = \sum_{k=1}^{n} \varphi(x_k, y_k)$$

for all $x_1, \ldots, x_n, y_1, \ldots, y_n \in A$. This functional extends to a continuous linear functional on A, which we also denote by τ. Of course,

$$\varphi(x, y) = \tau(xy) \quad (x, y \in A),$$

so (i) holds. $\square$

Corollary 4.10 *Let A be a unital nonassociative Banach algebra. The following conditions are equivalent:*

(i) *A is zpd.*

(ii) *$z \otimes w - zw \otimes 1 \in \overline{\mathrm{span}} \{x \otimes y \in A \widehat{\otimes} A \mid xy = 0\}$ for all $z, w \in A$.*

(iii) *If a continuous bilinear functional $\varphi \colon A \times A \to \mathbb{C}$ has the property that $\varphi(x, y) = 0$ whenever $x, y \in A$ are such that $xy = 0$, then*

$$\varphi(x, y) = \varphi(xy, 1) \quad (x, y \in A).$$

Proof (i) $\Longrightarrow$ (ii). Since

$$z \otimes w - zw \otimes 1 \in \ker \pi,$$

Proposition 4.9 (iii) gives the desired conclusion.

(ii) $\Longrightarrow$ (iii). Let φ be as in (iii). Then φ induces a continuous linear functional $\widehat{\varphi}$ on $A \widehat{\otimes} A$ such that

$$\widehat{\varphi}(x \otimes y) = \varphi(x, y)$$

and $\widehat{\varphi}$ vanishes on $\overline{\mathrm{span}} \{u \otimes v \in A \widehat{\otimes} A \mid uv = 0\}$. Hence, by (ii),

$$\varphi(x, y) - \varphi(xy, 1) = \widehat{\varphi}(x \otimes y - xy \otimes 1) = 0$$

for all $x, y \in A$.

(iii) $\Longrightarrow$ (i). Clear. $\square$

Providing various examples of zpd (nonassociative) Banach algebras is one of the goals of this book. We will deal with this later. Of course, the algebraic results from Part I already give some information. For example, an immediate consequence of Theorem 2.27 is that a finite-dimensional unital Banach algebra A is zpd if and only if A is generated by idempotents. In infinite dimensions, however, the results in the analytic context will be substantially different.

Modifying the arguments from Sect. 1.3 one can obtain some results on the stability of the zpd property in the analytic context. However, we omit this at this point, since the proofs will run more smoothly (in Sect. 5.2) after defining what we call property $\mathbb{B}$ (in Sect. 5.1). It should be mentioned that the latter concerns only (associative) Banach algebras, so we shall not consider stability in general nonassociative Banach algebras (this topic is perhaps not so interesting in view of the theory developed in the next chapters).

4.3   Point Derivations

Our aim now is to point out a condition that turns out to be one of the main obstacles for a nonassociative Banach algebra A to be zpd. Recall that Ω_A stands for the character space of A.

Definition 4.11 Let $\omega \in \Omega_A$. A linear functional $D: A \to \mathbb{C}$ is a *point derivation at* ω if

$$D(xy) = D(x)\omega(y) + \omega(x)D(y) \quad (x, y \in A).$$

Here is a simple example.

Example 4.12 Let $C^1[0, 1]$ be the algebra of continuously differentiable functions on $[0, 1]$. Endowed with the norm $\|f\| = \|f\|_\infty + \|f'\|_\infty$, where $\| \cdot \|_\infty$ is the sup-norm, it becomes a Banach algebra. Take $c \in [0, 1]$ and define ω by $\omega(f) = f(c)$. Then

$$D(f) = f'(c)$$

defines a continuous point derivation at ω.

Like Theorem 4.2, the following theorem also uses the (mild) assumption that A is third power-associative.

Theorem 4.13 *Let A be a third power-associative (nonassociative) Banach algebra. If there exists a nonzero continuous point derivation on A, then A is not a zpd Banach algebra.*

Proof Suppose D is a nonzero continuous point derivation at $\omega \in \Omega_A$. Define a continuous bilinear functional $\varphi: A \times A \to \mathbb{C}$ by

$$\varphi(x, y) = D(x)\omega(y).$$

If $x, y \in A$ are such that $xy = 0$, then

$$0 = D(xy) = D(x)\omega(y) + \omega(x)D(y),$$

and multiplying by $\omega(y)$ we arrive at

$$\begin{aligned}
0 &= D(x)\omega(y)^2 + \omega(x)\omega(y)D(y) \\
&= D(x)\omega(y)^2 + \omega(xy)D(y) \\
&= D(x)\omega(y)^2.
\end{aligned}$$

This implies that either $D(x) = 0$ or $\omega(y) = 0$, so that $\varphi(x, y) = 0$. Since A is zpd, it follows that there exists a (continuous) linear functional τ on A such that

$$\varphi(x, y) = \tau(xy) \quad (x, y \in A).$$

The kernels of D and ω are proper subspaces of A, so their union is not the whole A. Therefore, there exists an $x \in A$ such that $\omega(x) \neq 0$ and $D(x) \neq 0$. Using the assumption that A is third power-associative, we see that

$$\varphi(x^2, x) - \varphi(x, x^2) = \tau(x^2 x - xx^2) = 0.$$

However,

$$\varphi(x^2, x) - \varphi(x, x^2) = D(x^2)\omega(x) - D(x)\omega(x^2)$$
$$= 2D(x)\omega(x)\omega(x) - D(x)\omega(x)^2$$
$$= D(x)\omega(x)^2$$
$$\neq 0,$$

a contradiction.   □

Example 4.14 In view of Example 4.12, $A = C^1[0, 1]$ therefore is not a zpd Banach algebra. From the above discussion we see that, for any $c \in [0, 1]$,

$$\varphi(f, g) = \omega(f'g), \quad \text{where } \omega(f) = f(c),$$

is an example of a continuous bilinear functional $\varphi \colon A \times A \to \mathbb{C}$ for which there is no $\tau \in A'$ such that $\varphi(f, g) = \tau(fg)$ for all $f, g \in A$, but satisfies $\varphi(f, g) = 0$ whenever $fg = 0$.

This example can be generalized as follows. Observe that $fg = 0$ implies $f'g + fg' = 0$, which yields

$$(f'g)^2 = -f'gfg' = 0$$

and hence $f'g = 0$. Similarly we see that this further implies $f'g' = 0$. Therefore, if a continuous bilinear functional φ on A is of the form

$$\varphi(f, g) = \tau_1(fg) + \tau_2(f'g) + \tau_3(f'g') \tag{4.4}$$

for some $\tau_1, \tau_2, \tau_3 \in A'$, then φ has the property that $\varphi(f, g) = 0$ whenever $fg = 0$. It turns out that the converse is also true, that is, if a continuous bilinear functional φ has the latter property, then φ is of the form (4.4). This was proved in [5]. See also [163] for a related result on $C^n[0, 1]$.

Zero Product Determined Banach Algebras

5

We now focus on (associative) Banach algebras. Our approach to zpd Banach algebras is based on the related notion of a Banach algebra with property $\mathbb{B}$. In Banach algebras having bounded (left) approximate identities, property $\mathbb{B}$ is just equivalent to the zpd property. However, in many ways it is technically more suitable. Using it, we will be able to show, in particular, that all C^*-algebras and all group algebras $L^1(G)$, where G is a locally compact group, are zpd Banach algebras. Also, we will rely on property $\mathbb{B}$ in the study of the stability of the zpd property under various constructions.

5.1 Property $\mathbb{B}$

Note that in an associative algebra A, a bilinear functional φ of the form $\varphi(x, y) = \tau(xy)$ for some linear functional τ satisfies $\varphi(xy, z) = \varphi(x, yz)$ for all $x, y, z \in A$; conversely, if A is unital and φ satisfies the latter condition, then $\varphi(x, y) = \tau(xy)$, where τ is defined by $\tau(x) = \varphi(x, 1)$. The following concept is thus just a modification of the notion of a zpd Banach algebra.

Definition 5.1 A Banach algebra A has *property* $\mathbb{B}$ if every continuous bilinear functional $\varphi \colon A \times A \to \mathbb{C}$ with the property that $\varphi(x, y) = 0$ whenever $x, y \in A$ are such that $xy = 0$ satisfies

$$\varphi(xy, z) = \varphi(x, yz) \quad (x, y, z \in A).$$

© The Author(s), under exclusive license to Springer Nature Switzerland AG 2021
M. Brešar, *Zero Product Determined Algebras*, Frontiers in Mathematics,
https://doi.org/10.1007/978-3-030-80242-4_5

Property $\mathbb{B}$ was introduced [6], following what was called property $\mathbb{A}$. Thus, there are no deeper reasons for the name "property $\mathbb{B}$." We are keeping it primarily because it appears in a number of papers following [6]. Let us also mention that a Banach algebra A is said to have property $\mathbb{A}$ if it satisfies the assumptions of Theorem 5.20 below. However, we will not consider this property explicitly in this book.

We recall a basic notion of the theory of tensor products of modules. Let A be an algebra, and let M and N be a right and a left A-module, respectively. A bilinear map $\varphi: M \times N \to X$, where X is a linear space, is said to be *balanced* if

$$\varphi(my, n) = \varphi(m, yn) \quad (m \in M, \ n \in N, \ y \in A).$$

Thus, our requirement in Definition 5.1 is that φ is balanced. We will use this term in what follows.

Property $\mathbb{B}$ can be characterized as follows.

Proposition 5.2 *Let A be a Banach algebra. The following conditions are equivalent:*

(i) *A has property $\mathbb{B}$.*
(ii) *$zw \otimes u - z \otimes wu \in \overline{\text{span}} \{x \otimes y \in A \widehat{\otimes} A \mid xy = 0\}$ for all $z, w, u \in A$.*
(iii) *If a continuous bilinear map $\varphi: A \times A \to X$, where X is a normed space, has the property that $\varphi(x, y) = 0$ whenever $x, y \in A$ are such that $xy = 0$, then φ is balanced.*

Proof (i) $\implies$ (ii). Suppose $\xi \in (A \widehat{\otimes} A)'$ vanishes on span $\{x \otimes y \in A \widehat{\otimes} A \mid xy = 0\}$. Then $\varphi: A \times A \to \mathbb{C}$ defined by

$$\varphi(x, y) = \xi(x \otimes y)$$

is a continuous bilinear functional with the property that $\varphi(x, y) = 0$ whenever $x, y \in A$ are such that $xy = 0$. Hence, by assumption, φ is balanced. For all $z, w, u \in A$, we thus have

$$\xi(zw \otimes u - z \otimes wu) = \varphi(zw, u) - \varphi(z, wu) = 0,$$

which proves (ii).

(ii) $\implies$ (iii). Let Y be the completion of X, and let φ be as in (iii). The induced continuous linear map $\widehat{\varphi}: A \widehat{\otimes} A \to Y$, given by

$$\widehat{\varphi}(x \otimes y) = \varphi(x, y),$$

then vanishes on $\overline{\text{span}}\,\{x \otimes y \in A\widehat{\otimes}A \mid xy = 0\}$. We thus have

$$\varphi(zw, u) - \varphi(z, wu) = \widehat{\varphi}(zw \otimes u - z \otimes wu) = 0$$

for all $z, w, u \in A$.

(iii) $\Longrightarrow$ (i). Obvious. $\qquad\square$

It is clear that the following implication holds for any Banach algebra A:

$$A \text{ is zpd} \Longrightarrow A \text{ has property } \mathbb{B}. \qquad (5.1)$$

The reverse implication does not hold in general.

Example 5.3 Any Banach algebra A satisfying $A^3 = \{0\}$ has property $\mathbb{B}$. Indeed, this is because $xy \cdot z = x \cdot yz = 0$ for all $x, y, z \in A$, and so

$$\varphi(xy, z) = 0 = \varphi(x, yz)$$

automatically holds for every bilinear functional φ on A, which vanishes on pairs of elements whose product is 0. Therefore, it suffices to find an example of a non-zpd Banach algebra A with $A^3 = \{0\}$.

To this end, take any non-zpd Banach algebra B (for example, a finite-dimensional unital Banach algebra that is not generated by idempotents (see Theorem 2.27) or the Banach algebra $C^1[0, 1]$ (see Example 4.12)). Let A consist of all strictly upper triangular 3×3 matrices with entries in B. Note that A is a closed subalgebra of the Banach algebra $M_3(B)$ and that $A^3 = \{0\}$.

By assumption, there exists a continuous bilinear functional φ on B such that $\varphi(x, y) = 0$ whenever $xy = 0$, but φ is not of the form $\varphi(x, y) = \tau(xy)$ for a continuous linear functional τ on B. Define $\Phi : A \times A \to \mathbb{C}$ by

$$\Phi\left(\begin{bmatrix} 0 & x & y \\ 0 & 0 & z \\ 0 & 0 & 0 \end{bmatrix}, \begin{bmatrix} 0 & u & v \\ 0 & 0 & w \\ 0 & 0 & 0 \end{bmatrix}\right) = \varphi(x, w).$$

Of course, Φ is a continuous bilinear functional. Since

$$\begin{bmatrix} 0 & x & y \\ 0 & 0 & z \\ 0 & 0 & 0 \end{bmatrix} \cdot \begin{bmatrix} 0 & u & v \\ 0 & 0 & w \\ 0 & 0 & 0 \end{bmatrix} = \begin{bmatrix} 0 & 0 & xw \\ 0 & 0 & 0 \\ 0 & 0 & 0 \end{bmatrix},$$

Φ vanishes on pairs of elements whose product is 0. If A was zpd, there would exist a continuous linear functional T on A such that

$$\varphi(x, w) = \Phi\left(\begin{bmatrix} 0 & x & y \\ 0 & 0 & z \\ 0 & 0 & 0 \end{bmatrix}, \begin{bmatrix} 0 & u & v \\ 0 & 0 & w \\ 0 & 0 & 0 \end{bmatrix}\right) = T\left(\begin{bmatrix} 0 & 0 & xw \\ 0 & 0 & 0 \\ 0 & 0 & 0 \end{bmatrix}\right),$$

which contradicts our assumption on φ. Therefore, A is not a zpd Banach algebra but has property $\mathbb{B}$.

The example just given is of some interest also for the algebraic theory. Indeed, it shows that in the definition of a zpd algebra, we cannot substitute the condition $\varphi(xy, z) = \varphi(x, yz)$ for the condition that $\varphi(x, y) = \tau(xy)$. The next example is of a different nature.

Example 5.4 Let $A = S_2(H)$ be the Banach algebra of the *second Schatten class operators* of an infinite-dimensional separable Hilbert space H. Then $\varphi(x, y) = \text{tr}(xy)$ (here, $\text{tr}(\cdot)$ stands for the trace) is a continuous bilinear functional that obviously vanishes on pairs of elements whose product is 0. However, there is no continuous linear functional τ on A such that $\varphi(x, y) = \tau(xy)$ for all $x, y \in A$. Thus, A is not a zpd Banach algebra. On the other hand, from Theorem 5.14 (and Example 5.15) below, we see that A has property $\mathbb{B}$.

Note, however, that the reverse implication in (5.1) does hold if A is unital, so the two conditions are then equivalent. We will show that the unitality can be replaced by a milder assumption that A has a bounded left approximate identity. Let us first recall the definition of this and some related notions.

Let A be a Banach algebra. A *left approximate identity* for A is a net $(e_\lambda)_{\lambda \in \Lambda}$ in A such that

$$\lim e_\lambda x = x$$

for every $x \in A$. A *right approximate identity* for A is defined similarly. An *approximate identity* for A is a net $(e_\lambda)_{\lambda \in \Lambda}$, which is both a left and a right approximate identity for A. A (left/right) approximate identity $(e_\lambda)_{\lambda \in \Lambda}$ is *bounded* by $M > 0$ if $\|e_\lambda\| \leq M$ for every $\lambda \in \Lambda$.

Proposition 5.5 *Let A be a Banach algebra with a bounded left approximate identity. Then A is zpd if and only if A has property $\mathbb{B}$.*

Proof It is enough to prove the "if" part.

Let $\varphi \colon A \times A \to \mathbb{C}$ be a continuous bilinear functional such that $\varphi(x, y) = 0$ whenever $xy = 0$. By assumption, φ is balanced. Let $(e_\lambda)_{\lambda \in \Lambda}$ be a left approximate identity with bound M, and let $x_1, \ldots, x_n, y_1, \ldots, y_n \in A$. For any $\lambda \in \Lambda$, we have

$$
\left| \sum_{k=1}^{n} \varphi(e_\lambda x_k, y_k) \right| = \left| \sum_{k=1}^{n} \varphi(e_\lambda, x_k y_k) \right|
$$

$$
= \left| \varphi \left(e_\lambda, \sum_{k=1}^{n} x_k y_k \right) \right|
$$

$$
\leq \|\varphi\| M \left\| \sum_{k=1}^{n} x_k y_k \right\|,
$$

and hence

$$
\left| \sum_{k=1}^{n} \varphi(x_k, y_k) \right| = \left| \lim_{\lambda \in \Lambda} \sum_{k=1}^{n} \varphi(e_\lambda x_k, y_k) \right|
$$

$$
\leq \|\varphi\| M \left\| \sum_{k=1}^{n} x_k y_k \right\|.
$$

By Proposition 4.9 (ii), A is zpd.   □

We will often assume that our Banach algebras are unital or have bounded (left) approximate identities. Thus, for the most part, we will consider property $\mathbb{B}$ just as a useful technical reinterpretation of the zpd condition.

For some further variations of the notion of a zpd Banach algebra, see the recent papers [80–82].

5.2   Stability Under Analytic Constructions

In this section, we present analytic analogues of algebraic results from Sect. 1.3 on the stability of the zpd property under various constructions. Unlike in the algebraic setting, however, we will not consider nonassociative Banach algebras. Our approach is based on property $\mathbb{B}$.

Theorem 5.6 *Let A be a Banach algebra with property $\mathbb{B}$, let B be any Banach algebra, and let $\theta \colon A \to B$ be a continuous homomorphism with dense range. Then B has property $\mathbb{B}$.*

Proof Let $\varphi \colon B \times B \to \mathbb{C}$ be a continuous bilinear functional as in Definition 5.1. We define $\psi \colon A \times A \to \mathbb{C}$ by

$$\psi(x, y) = \varphi\big(\theta(x), \theta(y)\big) \quad (x, y \in A).$$

Note that ψ is a continuous bilinear functional with the property that $\psi(x, y) = 0$ whenever $xy = 0$. Hence, ψ is balanced, which can be written as

$$\varphi\big(\theta(x)\theta(y), \theta(z)\big) = \varphi\big(\theta(x), \theta(y)\theta(z)\big) \quad (x, y, z \in A).$$

Since the range of θ is dense, it follows that $\varphi(uv, w) = \varphi(u, vw)$ for all $u, v, w \in B$; that is, φ is balanced. $\qquad\square$

Corollary 5.7 *Let A be a zpd Banach algebra with a bounded left approximate identity, let B be any Banach algebra, and let $\theta \colon A \to B$ be a continuous homomorphism with dense range. Then B is zpd.*

Proof Since A is zpd, it also has property $\mathbb{B}$. Therefore, Theorem 5.6 implies that B has property $\mathbb{B}$ as well. On account of Proposition 5.5, it is enough to prove that B has a bounded left approximate identity. Let $(e_\lambda)_{\lambda \in \Lambda}$ be a left approximate identity for A with bound $M > 0$. We claim that $(\theta(e_\lambda))_{\lambda \in \Lambda}$ is a bounded left approximate identity for B. Of course,

$$\|\theta(e_\lambda)\| \le \|\theta\| M \quad (\lambda \in \Lambda).$$

Take $u \in B$, and let $\varepsilon > 0$. Since θ has dense range, there exists an $x \in A$ such that $\|u - \theta(x)\| < \varepsilon$, and, since $(e_\lambda)_{\lambda \in \Lambda}$ is a left approximate identity for A, there exists a $\nu \in \Lambda$ such that $\|x - e_\lambda x\| < \varepsilon$ whenever $\lambda \ge \nu$. Then, for each $\lambda \in \Lambda$ with $\lambda \ge \nu$, we have

$$\|u - \theta(e_\lambda)u\|$$
$$\le \|u - \theta(x)\| + \|\theta(x) - \theta(e_\lambda)\theta(x)\| + \|\theta(e_\lambda)\theta(x) - \theta(e_\lambda)u\|$$
$$\le \|u - \theta(x)\| + \|\theta\| \|x - e_\lambda x\| + \|\theta(e_\lambda)\| \|\theta(x) - u\|$$
$$< (1 + \|\theta\| + \|\theta\| M)\varepsilon.$$

This shows that $\lim_{\lambda \in \Lambda} \theta(e_\lambda)u = u$. $\qquad\square$

Theorem 5.8 *Let A be a Banach algebra with property* $\mathbb{B}$, *and let I be a closed ideal of A. Then,*

(a) *A/I has property* $\mathbb{B}$.
(b) *If* $\overline{AI} = \overline{IA} = I$, *then I has property* $\mathbb{B}$.

Proof

(a) Apply Theorem 5.6 to the quotient map from A onto A/I.
(b) Take a continuous bilinear functional $\varphi \colon I \times I \to \mathbb{C}$ as in Definition 5.1. Pick $u, v \in I$, and define $\psi_{u,v} \colon A \times A \to \mathbb{C}$ by

$$\psi_{u,v}(x, y) = \varphi(ux, yv).$$

It is clear that $\psi_{u,v}$ is a continuous bilinear functional such that $\psi_{u,v}(x, y) = 0$ whenever $xy = 0$. Since A has property $\mathbb{B}$, it follows that $\psi_{u,v}$ is balanced, so that

$$\varphi(uxy, zv) = \varphi(ux, yzv) \quad (x, y, z \in A, u, v \in I).$$

We thus get $\varphi(sy, t) = \varphi(s, yt)$ for all $s \in IA$, $y \in A$, and $t \in AI$. Using the assumption that $\overline{AI} = \overline{IA} = I$ along with the continuity of φ, we see that φ is balanced.   □

In the next proof, we will use *Cohen's factorization theorem* , which states that if A has a bounded left approximate identity, then every $x \in A$ can be written as $x = yz$ for some $y, z \in A$ (see, e.g., [76, Corollary 2.9.25]).

Corollary 5.9 *Let A be a zpd Banach algebra, and let I be a closed ideal of A.*

(a) *If A has a bounded left approximate identity, then A/I is zpd.*
(b) *If I has a bounded left approximate identity, then I is zpd.*

Proof

(a) Apply Corollary 5.7 to the quotient map from A onto A/I.
(b) Since A is zpd, it has property $\mathbb{B}$. Cohen's factorization theorem implies that $I^2 = I$, so Theorem 5.8 (b) shows that I has property $\mathbb{B}$, and hence I is zpd by Proposition 5.5.   □

Let $(A_i)_{i \in I}$ be a family of Banach algebras, and take $1 \leq p < \infty$. Define

$$\ell^p(I, A_i) = \left\{ (x_i)_{i \in I} \in \prod_{i \in I} A_i \mid \|(x_i)_{i \in I}\| = \left(\sum_{i \in I} \|x_i\|^p \right)^{1/p} < \infty \right\},$$

$$\ell^\infty(I, A_i) = \left\{ (x_i)_{i \in I} \in \prod_{i \in I} A_i \mid \|(x_i)_{i \in I}\| = \sup_{i \in I} \|x_i\| < \infty \right\}.$$

Then $\ell^p(I, A_i)$ and $\ell^\infty(I, A_i)$ are Banach spaces. The space $\bigoplus_{i \in I} A_i$ (defined in Sect. 1.3) is dense in $\ell^p(I, A_i)$ for each $1 \leq p < \infty$; its closure in $\ell^\infty(I, A_i)$ is denoted by $c_0(I, A_i)$. The Banach spaces $\ell^p(I, A_i)$, $1 \leq p \leq \infty$, and $c_0(I, A_i)$ are Banach algebras with respect to the coordinatewise product.

Theorem 5.10 *Let $(A_i)_{i \in I}$ be a family of Banach algebras. Then each of the Banach algebras $\ell^p(I, A_i)$, with $1 \leq p < \infty$, and $c_0(I, A_i)$ has property $\mathbb{B}$ if and only if each of the Banach algebras A_i has property $\mathbb{B}$.*

Proof Let A denote either $\ell^p(I, A_i)$ or $c_0(I, A_i)$. For each $i \in I$, let $\iota_i : A_i \to A$ be the natural embedding, and let $\theta_i : A \to A_i$ be the natural projection.

Suppose that A has property $\mathbb{B}$. Then, for each $i \in I$, we can apply Theorem 5.6 to the map θ_i to obtain that A_i has property $\mathbb{B}$.

We now assume that A_i has property $\mathbb{B}$ for each $i \in I$. Let $\varphi : A \times A \to \mathbb{C}$ be a continuous bilinear functional such that $\varphi(x, y) = 0$ whenever $xy = 0$. For each $i \in I$, consider the map $\varphi_i : A_i \times A_i \to \mathbb{C}$ given by

$$\varphi_i(x_i, y_i) = \varphi\big(\iota_i(x_i), \iota_i(y_i)\big) \quad (x_i, y_i \in A_i).$$

It is clear that φ_i is a continuous bilinear functional with the property that $\varphi_i(x_i, y_i) = 0$ whenever $x_i y_i = 0$. Since A_i has property $\mathbb{B}$, we conclude that φ_i is balanced, so that

$$\varphi(\iota_i(x_i)\iota_i(y_i), \iota_i(z_i)) = \varphi(\iota_i(x_i), \iota_i(y_i)\iota_i(z_i)) \quad (x_i, y_i, z_i \in A_i).$$

This clearly implies that $\varphi(xy, z) = \varphi(x, yz)$ for all $x, y, z \in \bigoplus_{i \in I} A_i$. Since $\bigoplus_{i \in I} A_i$ is dense in A and φ is continuous, we arrive at the desired conclusion that $\varphi(xy, z) = \varphi(x, yz)$ for all $x, y, z \in A$. $\qquad\square$

Corollary 5.11 *Let $(A_i)_{i \in I}$ be a family of zpd Banach algebras each of which has a left approximate identity with bound M. Then $c_0(I, A_i)$ is zpd if and only if each of the Banach algebras A_i is zpd.*

Proof Suppose $A = c_0(I, A_i)$ is zpd, so that A has property $\mathbb{B}$ as well. Let $i \in I$. Then Theorem 5.10 tells us that A_i has property $\mathbb{B}$, and, since A_i has a bounded left approximate identity, Proposition 5.5 implies that A_i is zpd.

Assume now that A_i is zpd for each $i \in I$. Then A_i has property $\mathbb{B}$ for each $i \in I$, and Theorem 5.10 shows that $c_0(I, A_i)$ has property $\mathbb{B}$. By Proposition 5.5, it suffices to prove that $c_0(I, A_i)$ has a bounded left approximate identity. Take $(x_i)_{i \in I} \in \bigoplus_{i \in I} A_i$, and let $\varepsilon > 0$. Set $I_0 = \{i \in I \mid x_i \neq 0\}$. Then I_0 is finite, and, for each $i \in I_0$, we choose $u_i \in A_i$ such that $\|u_i\| \leq M$ and $\|x_i - u_i x_i\| < \varepsilon$. For $i \in I \setminus I_0$, we take $u_i = 0$. Then $(u_i)_{i \in I} \in c_0(I, A_i)$, $\|(u_i)_{i \in I}\| \leq M$, and

$$\|(x_i)_{i \in I} - (u_i)_{i \in I}(x_i)_{i \in I}\| < \varepsilon.$$

Since $\bigoplus_{i \in I} A_i$ is dense in $c_0(I, A_i)$, [76, Proposition 2.9.14 (ii)] implies that $c_0(I, A_i)$ has a left approximate identity of bound M, which proves our claim. $\qquad\square$

Theorem 5.12 *If Banach algebras A and B have property $\mathbb{B}$, then so does $A \widehat{\otimes} B$.*

Proof Let $\varphi \colon (A \widehat{\otimes} B) \times (A \widehat{\otimes} B) \to \mathbb{C}$ be a continuous bilinear functional as in Definition 5.1. For $u, v \in B$, we define $\varphi_{u,v} \colon A \times A \to \mathbb{C}$ by

$$\varphi_{u,v}(x, y) = \varphi(x \otimes u, y \otimes v) \quad (x, y \in A).$$

Then $\varphi_{u,v}$ is a continuous bilinear functional with the property that $\varphi_{u,v}(x, y) = 0$ whenever $xy = 0$. As A has property $\mathbb{B}$, it follows that

$$\varphi_{u,v}(xy, z) = \varphi_{u,v}(x, yz) \quad (x, y, z \in A);$$

that is,

$$\varphi(xy \otimes u, z \otimes v) = \varphi(x \otimes u, yz \otimes v) \quad (x, y, z \in A, \ u, v \in B).$$

Likewise, since B has property $\mathbb{B}$,

$$\varphi(x \otimes uv, y \otimes w) = \varphi(x \otimes u, y \otimes vw) \quad (x, y \in A, \ u, v, w \in B).$$

The last two identities imply that for all $x, y, z \in A$ and $u, v, w \in B$,

$$\varphi((x \otimes u)(y \otimes v), z \otimes w)$$
$$= \varphi(xy \otimes uv, z \otimes w)$$
$$= \varphi(x \otimes uv, yz \otimes w)$$

$$=\varphi(x \otimes u, \, yz \otimes vw)$$

$$=\varphi(x \otimes u, \, (y \otimes v)(z \otimes w)).$$

Since the linear span of simple tensors $x \otimes u$, with $x \in A$ and $u \in B$, is dense in $A \widehat{\otimes} B$ and φ is continuous, it may be concluded that φ is balanced. $\square$

Corollary 5.13 *Let A and B be zpd Banach algebras with bounded left approximate identities. Then $A \widehat{\otimes} B$ is zpd.*

Proof Since both A and B have property $\mathbb{B}$, Theorem 5.12 tells us that $A \widehat{\otimes} B$ has property $\mathbb{B}$, and since both A and B have bounded left approximate identities, [76, Proposition 2.9.21] tells us that $A \widehat{\otimes} B$ has a bounded left approximate identity. Thus, $A \widehat{\otimes} B$ is zpd by Proposition 5.5. $\square$

5.3   Examples and Non-examples of zpd Banach Algebras

This section is of crucial importance for this book. We will show that the class of zpd Banach algebras is large, but not exhaustive. Notably, it includes all C^*-algebras and all group algebras $L^1(G)$, but, on the other hand, it does not include Banach algebras of continuous linear operators $\mathscr{B}(X)$ for all Banach spaces X.

Let us start by recording a simple result analogous to Theorem 2.15.

Theorem 5.14 *If a Banach algebra A is generated by idempotents, then A has property $\mathbb{B}$ (and is hence zpd if it has a left approximate identity).*

Proof Let $\varphi : A \times A \to \mathbb{C}$ be a continuous bilinear functional such that $\varphi(x, y) = 0$ whenever $xy = 0$. We first repeat the argument from the proof of Lemma 2.2. For any idempotent $e \in A$ and any $x, z \in A$, we have

$$xe \cdot (z - ez) = 0 \text{ and } (x - xe) \cdot ez = 0.$$

Hence,

$$\varphi(xe, z - ez) = 0 \text{ and } \varphi(x - xe, ez) = 0,$$

which we can rewrite as

$$\varphi(xe, z) = \varphi(xe, ez) \text{ and } \varphi(x, ez) = \varphi(xe, ez).$$

Comparing these two identities, we obtain $\varphi(xe, z) = \varphi(x, ez)$. Thus, the set

$$A_0 = \{r \in A \mid \varphi(xr, z) = \varphi(x, rz) \text{ for all } x, z \in A\}$$

contains all idempotents in A. Observe that A_0 is a subalgebra of A. Moreover, it is closed since φ is continuous. As A is generated (as a Banach algebra) by idempotents, it follows that $A_0 = A$, meaning that φ is balanced. Thus, A has property $\mathbb{B}$ (which is, by Proposition 5.5, equivalent to A being zpd if A has a left approximate identity).   □

Example 5.15 If A is any unital Banach algebra, then $M_n(A)$, $n \geq 2$, is a Banach algebra generated by idempotents (see Corollary 2.4). Also, the triangular algebras $\text{Tri}(R_1, M, R_2)$ from Corollary 2.5 may be Banach algebras for appropriate choices of R_1, R_2, and M. As another example, take a Banach space X, and let $\mathscr{F}(X)$ be the algebra of continuous finite-rank linear operators on X. Then $\mathscr{F}(X)$ is generated by idempotents (in the algebraic sense), and so its operator norm closure in $\mathscr{B}(X)$ is a Banach algebra generated by idempotents. It is called the algebra of *approximable operators* on X and is denoted by $\mathscr{A}(X)$. Onc can take closures with respect to some other norms on $\mathscr{F}(X)$ and in this way obtain further examples (e.g., the *Schatten class operators* $S_p(H)$, where H is a Hilbert space).

We could list more examples of Banach algebras that are generated by idempotents (see [147, Chapter 3] for a thorough discussion on such algebras). This class of Banach algebras indeed includes some important examples, but nevertheless it is somewhat narrow; it certainly does not include all C^*-algebras and all group algebras $L^1(G)$. We thus need a different approach to cover these algebras. Instead of with idempotents, we will deal with the elements from the following definition.

Definition 5.16 An invertible element u of a unital Banach algebra is said to be *doubly power-bounded* if

$$\sup_{k \in \mathbb{Z}} \|u^k\| < \infty.$$

We remark that if e is an idempotent, then $u = 1 - 2e$ satisfies $u^2 = 1$ and is hence doubly power-bounded. Using this, it is easy to see that the results that follow also cover Theorem 5.14 (if A is not unital, then one has to involve its unitization in the proof). Our approach in the analytic context can be therefore viewed as a generalization of the algebraic approach from Part I.

Our goal now is to establish the basic auxiliary result, Lemma 5.18. Its proof uses some standard concepts and results from the commutative Banach algebra theory, which we will now present in a succinct manner, focusing solely on what is needed for our proof. For details, we refer the reader to the books [76, 110], which consider these topics in their general contexts.

For any compact space X, we write $C(X)$ for the algebra of all continuous functions from X to $\mathbb{C}$. We will actually consider only the case where X is either the *circle group*

$$\mathbb{T} = \{z \in \mathbb{C} \mid |z| = 1\}$$

or the direct product $\mathbb{T} \times \mathbb{T}$.

By $\widehat{f}(k)$, we denote the kth Fourier coefficient of $f \in C(\mathbb{T})$, i.e.,

$$\widehat{f}(k) = \frac{1}{2\pi} \int_{-\pi}^{\pi} f(e^{it}) e^{-ikt} dt.$$

It is well-known that

$$A(\mathbb{T}) = \{ f \in C(\mathbb{T}) \mid \sum_{k=-\infty}^{\infty} |\widehat{f}(k)| < \infty \}$$

is a commutative unital Banach algebra under the pointwise multiplication and the norm given by

$$\|f\|_{A(\mathbb{T})} = \sum_{k=-\infty}^{\infty} |\widehat{f}(k)|.$$

We call $A(\mathbb{T})$ the *Fourier algebra of* $\mathbb{T}$. We also need the *Fourier algebra of* $\mathbb{T} \times \mathbb{T}$,

$$A(\mathbb{T} \times \mathbb{T}) = \{ F \in C(\mathbb{T} \times \mathbb{T}) \mid \sum_{j=-\infty}^{\infty} \sum_{k=-\infty}^{\infty} |\widehat{F}(j, k)| < \infty \},$$

where

$$\widehat{F}(j, k) = \frac{1}{4\pi^2} \int_{-\pi}^{\pi} \int_{-\pi}^{\pi} F(e^{is}, e^{it}) e^{-ijs} e^{-ikt} ds\, dt.$$

Of course, $A(\mathbb{T} \times \mathbb{T})$ is also a commutative unital Banach algebra with the norm

$$\|F\|_{A(\mathbb{T} \times \mathbb{T})} = \sum_{j=-\infty}^{\infty} \sum_{k=-\infty}^{\infty} |\widehat{F}(j, k)|.$$

For any functions $f, g \in A(\mathbb{T})$, we write $f \otimes g$ for the function on $\mathbb{T} \times \mathbb{T}$ given by

$$(f \otimes g)(z, w) = f(z)g(w).$$

From

$$(\widehat{f \otimes g})(j, k) = \widehat{f}(j)\widehat{g}(k),$$

we see that $f \otimes g \in A(\mathbb{T} \times \mathbb{T})$ (moreover, $\|f \otimes g\|_{A(\mathbb{T} \times \mathbb{T})} = \|f\|_{A(\mathbb{T})}\|g\|_{A(\mathbb{T})}$).

In the next paragraphs, let G stand for either $\mathbb{T}$ or $\mathbb{T} \times \mathbb{T}$ (as a matter of fact, we could take any Abelian locally compact group for G, but we have no need for this generality).

The first fact that we need is that the algebra $A(G)$ is *regular* [76, Theorem 4.5.19], meaning that for every nonempty closed subset S of G and every $t \in G \setminus S$, there exists an $f \in A(G)$ such that $f(t) = 1$ and $f|_S = 0$.

Let J be a closed ideal of $A(G)$. We call

$$\mathfrak{h}(J) = \{g \in G \mid F(g) = 0 \text{ for all } F \in J\}$$

the *hull of* J. Obviously, $\mathfrak{h}(J)$ is a closed subset of G. Now let S be any nonempty closed subset of G. Observe that

$$I = \{F \in A(G) \mid F|_S = 0\} \tag{5.2}$$

is a closed ideal of A, and, by the regularity of $A(G)$, $\mathfrak{h}(I) = S$. If there are no other closed ideals whose hull is equal to S, then we say that S is a *set of synthesis* for $A(G)$. Thus, more precisely, a nonempty closed subset S of G is a set of synthesis for $A(G)$ if, for any closed ideal I of A, $\mathfrak{h}(I) = S$ implies that (5.2) holds. An important fact for us is that every closed subgroup of G is a set of synthesis for $A(G)$—see [110, Theorem 5.5.8] or [111, Theorem 6.1.9]. We will actually need only a special case where $G = \mathbb{T} \times \mathbb{T}$, and the closed subgroup is

$$\Delta = \{(z, z) \mid z \in \mathbb{T}\}.$$

Thus, the following lemma will be relevant for us.

Lemma 5.17 Δ *is a set of synthesis for* $A(\mathbb{T} \times \mathbb{T})$.

We are now in a position to establish our basic lemma.

Lemma 5.18 *Let A be a Banach algebra, and let $\varphi : A \times A \to \mathbb{C}$ be a continuous bilinear functional satisfying $\varphi(x, y) = 0$ whenever $xy = 0$. Suppose A is an ideal of a unital Banach algebra B. Then $\varphi(xu, y) = \varphi(x, uy)$ for all $x, y \in A$ and all doubly power-bounded elements $u \in B$.*

Proof We fix x, y, and u. Define $\theta : A(\mathbb{T}) \to B$ by

$$\theta(f) = \sum_{j=-\infty}^{\infty} \widehat{f}(j) u^j.$$

Note that θ is a continuous algebra homomorphism. Next, define

$$\phi : A(\mathbb{T}) \times A(\mathbb{T}) \to \mathbb{C}$$

by

$$\phi(f, g) = \varphi\big(x\theta(f), \theta(g)y\big),$$

and, finally, define

$$\Phi : A(\mathbb{T} \times \mathbb{T}) \to \mathbb{C}$$

by

$$\Phi(F) = \sum_{j=-\infty}^{\infty} \sum_{k=-\infty}^{\infty} \widehat{F}(j, k) \phi(z^j, z^k).$$

Observe that ϕ is a continuous bilinear functional, and hence Φ is a continuous linear functional.

Recall that Δ stands for $\{(z, z) \,|\, z \in \mathbb{T}\}$. We claim that it is enough to prove the following:

($\star$) If $F \in A(\mathbb{T} \times \mathbb{T})$ vanishes on Δ, then $\Phi(F) = 0$.

Indeed, applying ($\star$) to the function

$$F(z, w) = z - w,$$

we obtain

$$0 = \Phi(F) = \phi(z, 1) - \phi(1, z) = \varphi(xu, y) - \varphi(x, uy),$$

which is the desired conclusion. The rest of the proof will be therefore devoted to establishing ($\star$).

Take $f, g \in A(\mathbb{T})$. We have

$$\Phi(f \otimes g) = \sum_{j=-\infty}^{\infty} \sum_{k=-\infty}^{\infty} \widehat{(f \otimes g)}(j, k)\phi(z^j, z^k)$$

$$= \sum_{j=-\infty}^{\infty} \sum_{k=-\infty}^{\infty} \widehat{f}(j)\widehat{g}(k)\phi(z^j, z^k)$$

$$= \phi\left(\sum_{j=-\infty}^{\infty} \widehat{f}(j)z^j, \sum_{k=-\infty}^{\infty} \widehat{g}(k)z^k\right)$$

$$= \phi(f, g).$$

Assume now that $fg = 0$. Then

$$x\theta(f) \cdot \theta(g)y = x\theta(fg)y = 0,$$

and so, by our assumption on φ,

$$\phi(f, g) = \varphi\big(x\theta(f), \theta(g)y\big) = 0.$$

By what we have just shown, this can be written as $\Phi(f \otimes g) = 0$. Thus, Φ vanishes on every function $f \otimes g$ with $fg = 0$. Since Φ is linear and continuous, it also vanishes on

$$I = \overline{\text{span}} \{f \otimes g \mid f, g \in A(\mathbb{T}) \text{ and } fg = 0\}.$$

We remark that I is an ideal of $A(\mathbb{T} \times \mathbb{T})$. Indeed, this follows from the fact that the linear span of the functions $z^j \otimes z^k$, $j, k \in \mathbb{Z}$, is dense in $A(\mathbb{T} \times \mathbb{T})$.

We now claim that it is enough to prove the following:

($\star\star$) $\mathfrak{h}(I) = \Delta$.

This is because Δ is, by Lemma 5.17, a set of synthesis for $A(\mathbb{T} \times \mathbb{T})$; therefore, ($\star\star$) implies that I consists of all functions in $A(\mathbb{T} \times \mathbb{T})$ that vanish on Δ, which along with $\Phi(I) = \{0\}$ yields ($\star$).

To start the proof of ($\star\star$), first observe that $f \otimes g$ vanishes on Δ whenever $f, g \in A(\mathbb{T})$ are such that $fg = 0$. Of course, the same is true for every linear combination of such functions $f \otimes g$. Since $|F(z, z)| \leq \|F\|_{A(\mathbb{T} \times \mathbb{T})}$ for every $F \in A(\mathbb{T} \times \mathbb{T})$ and every $z \in \mathbb{T}$, it follows that every function from I also vanishes on Δ. This means that $\Delta \subseteq \mathfrak{h}(I)$.

To prove the converse inclusion, take distinct $z, w \in \mathbb{T}$. Choose an open neighborhood U of z and an open neighborhood V of w such that $U \cap V = \emptyset$. Since $A(\mathbb{T})$ is regular, there exist functions $f, g \in A(\mathbb{T})$ satisfying $f(z) = g(w) = 1$, f is zero on the complement of U, and g is zero on the complement of V. In particular, $fg = 0$, and so $f \otimes g \in I$. From

$$(f \otimes g)(z, w) = f(z)g(w) = 1,$$

we thus see that $(z, w) \notin \mathfrak{h}(I)$. We have thereby shown that $\mathfrak{h}(I) \subseteq \Delta$, which completes the proof of (⋆⋆).   □

Recall that a *C*-algebra* is a Banach algebra A with involution ∗ such that $\|x^*x\| = \|x\|^2$ for all $x \in A$ (here, by an *involution* , we mean a map $x \mapsto x^*$ from A to A satisfying $(x + y)^* = x^* + y^*$, $(\lambda x)^* = \bar{\lambda} x^*$, $(xy)^* = y^* x^*$, and $(x^*)^* = x$ for all $x, y \in A$ and $\lambda \in \mathbb{C}$). Assuming that A is unital, we say that $u \in A$ is *unitary* if $u^* = u^{-1}$. These elements are obviously doubly power-bounded. It is a standard fact that every element in a unital C^*-algebra is a linear combination of four unitaries.

We are now ready to state our first application of Lemma 5.18.

Theorem 5.19 *Every C*-algebra is a zpd Banach algebra.*

Proof We set $B = A$ if A is unital, and if not, then let B be the unitization of A (i.e., the C^*-algebra obtained by adjoining a unity to A). In any case, A is an ideal of B and B is a unital C^*-algebra. Since, as just mentioned, unitaries linearly span B and are doubly power-bounded, Lemma 5.18 shows that A has property $\mathbb{B}$. As every C^*-algebra has a bounded approximate identity [76, Theorem 3.2.21], A is zpd by Proposition 5.5.   □

In order to cover group algebras $L^1(G)$, we will have to apply Lemma 5.18 in a more subtle manner. Let us first recall a few more notions from the Banach algebra theory.

Let A be a Banach algebra. If linear maps $L, R : A \to A$ satisfy

$$L(xy) = L(x)y, \quad R(xy) = xR(y), \quad xL(y) = R(x)y \quad (x, y \in A),$$

then we call the pair (L, R) a *multiplier* on A. For example, (L_a, R_a) is a multiplier for every $a \in A$, where L_a and R_a are left and right multiplication operators. Denote the set of all multipliers on A by $\mathscr{M}(A)$.

Assume from now on that A is both *left faithful* and *right faithful*, meaning that, for any $a \in A$, each of the conditions $aA = \{0\}$ and $Aa = \{0\}$ implies $a = 0$. Then $\mathscr{M}(A)$ is easily seen to be a closed unital subalgebra of $\mathscr{B}(A) \times \mathscr{B}(A)^{\mathrm{op}}$ [76, Proposition 2.5.12 (i)]; here, $\mathscr{B}(A)$ stands for the Banach algebra of all bounded linear operators on A and $\mathscr{B}(A)^{\mathrm{op}}$ for the opposite algebra of $\mathscr{B}(A)$. We call $\mathscr{M}(A)$ the *multiplier algebra* of A. Next we remark that we can continuously embed A into $\mathscr{M}(A)$ by $a \mapsto (L_a, R_a)$ [76, Proposition 2.5.12 (ii)]. From now on, we identify A with its image via this embedding. It is easy to see that A is an ideal of $\mathscr{M}(A)$ (and hence $A = \mathscr{M}(A)$ if A is unital).

The *strong operator topology* on $\mathcal{M}(A)$, denoted by so, is defined by the family of seminorms $\{p_x \mid x \in A\}$, where

$$p_x\big((L, R)\big) = \max\big\{\|L(x)\|, \|R(x)\|\big\} \quad \big((L, R) \in \mathcal{M}(A)\big).$$

Thus, a net $(\mu_i)_{i \in I}$ in $\mathcal{M}(A)$ converges to $\mu \in \mathcal{M}(A)$ with respect to the strong operator topology if, for every $x \in A$,

$$\lim_{i \in I} \mu_i x = \mu x \quad \text{and} \quad \lim_{i \in I} x \mu_i = x \mu$$

with respect to the norm topology.

Denote by $\mathscr{D}(A)$ the set of all doubly power-bounded elements in $\mathcal{M}(A)$ and by alg $\mathscr{D}(A)$ the subalgebra of $\mathcal{M}(A)$ generated by $\mathscr{D}(A)$. We will be concerned with $\overline{\text{alg } \mathscr{D}(A)}^{\text{so}}$, the closure of alg $\mathscr{D}(A)$ in $\mathcal{M}(A)$ with respect to the strong operator topology.

Theorem 5.20 *Let A be a left and right faithful Banach algebra. If $A \subseteq \overline{\text{alg } \mathscr{D}(A)}^{\text{so}}$, then A has property $\mathbb{B}$.*

Proof Let $\varphi : A \times A \to \mathbb{C}$ be a continuous bilinear functional. Observe that the set

$$\mathscr{A} = \{\mu \in \mathcal{M}(A) \mid \varphi(x\mu, y) = \varphi(x, \mu y) \text{ for all } x, y \in A\}$$

is a subalgebra of $\mathcal{M}(A)$. We claim that $\mathscr{A}$ is closed in the strong operator theory. Indeed, let $(\mu_i)_{i \in I}$ be a net in $\mathscr{A}$ converging to μ in the strong operator topology. As $\lim_{i \in I} x \mu_i = x\mu$ and $\lim_{i \in I} \mu_i y = \mu y$, it follows that for any $x, y \in A$,

$$\varphi(x\mu, y) = \lim_{i \in I} \varphi(x\mu_i, y)$$

$$= \lim_{i \in I} \varphi(x, \mu_i y)$$

$$= \varphi(x, \mu y),$$

which proves our claim.

Assume now that φ satisfies the condition that $\varphi(x, y) = 0$ whenever $xy = 0$. Lemma 5.18 then tells us that $\mathscr{A}$ contains all doubly power-bounded elements in $\mathcal{M}(A)$. Therefore,

$$A \subseteq \overline{\text{alg } \mathscr{D}(A)}^{\text{so}} \subseteq \mathscr{A},$$

which shows that φ is balanced. $\square$

Now, let G be a locally compact group, and let A be the *group algebra* $L^1(G)$ (the definition and basic properties can be found in [76] and other standard text). The multiplier algebra $\mathcal{M}(A)$ can be then identified with the Banach algebra $M(G)$ of all complex-valued regular Borel measures on G [76, Theorem 3.3.40]. Furthermore, for any $\in G$, the unit point mass measure δ_s is a doubly power-bounded element of $M(G)$, and, according to [76, Proposition 3.3.41],

$$M(G) = \overline{\mathrm{span}\,\{\delta_s \mid s \in G\}}^{\mathrm{so}}.$$

Therefore, A satisfies the conditions of Theorem 5.20. Since it has a bounded approximate identity [76, Theorem 3.3.23], the following theorem holds.

Theorem 5.21 *Let G be a locally compact group. Then $L^1(G)$ is a zpd Banach algebra.*

We conclude this section with a few non-examples of zpd Banach algebras.

Example 5.22 The zpd property is based on zero-divisors. Therefore, Banach algebras such as the disc algebra certainly cannot be zpd (see Corollary 1.8 and Corollary 4.10).

Example 5.23 For finite-dimensional Banach algebras, the notions of "zpd algebra" and "zpd Banach algebra" of course coincide. Therefore, a finite-dimensional unital Banach algebra A that is not generated by idempotents is not a zpd Banach algebra by Theorem 2.27.

Example 5.24 As noticed in Example 4.14, the Banach algebra $C^1[0, 1]$ has nonzero continuous point derivations and hence is not zpd by Theorem 4.13. Of course, the same is true for $C^n[0, 1]$ for any $n \geq 1$.

Example 5.25 If G is a nondiscrete locally compact abelian group, then there are nonzero continuous point derivations on $M(G)$ [63]. Therefore, $M(G)$ is not zpd.

Example 5.26 Let X be a Banach space. It is safe to say that the Banach algebra $\mathcal{B}(X)$ of all continuous linear operators on X is often zpd. Indeed, if X is isomorphic to Y^n for some Banach space Y, then $\mathcal{B}(X)$ is isomorphic to $M_n(\mathcal{B}(Y))$ and is therefore zpd (see Example 5.15). However, there also exist Banach spaces X such that $\mathcal{B}(X)$ is not zpd. Indeed, [145] provides an example of a Banach space X such that $\mathcal{B}(X)$ contains an ideal M that has codimension 1 and satisfies $\overline{M^2} \neq M$. Observe that the linear functional ω on $\mathcal{B}(X)$ whose kernel is M and sends 1 to 1 is a character. Take a nonzero continuous linear functional D on $\mathcal{B}(X)$ such that $D(1) = 0$ and $D(\overline{M^2}) = \{0\}$. Note that $\omega(x)1 - x \in M$

for every $x \in \mathcal{B}(X)$. Hence,

$$D\big((\omega(x)1 - x)(\omega(y)1 - y)\big) \in D(M^2) = \{0\},$$

from which we infer that D is a point derivation at ω. Therefore, $\mathcal{B}(X)$ is not a zpd Banach algebra.

Most of the results from this chapter are taken from [6]. It is noteworthy that the proof of Lemma 5.18 is simpler than the one given in that paper. It was suggested to the author by Armando Villena.

Zero Lie/Jordan Product Determined Banach Algebras

6

Our approach to zLpd (zero Lie product determined) Banach algebras and zJpd (zero Jordan product determined) Banach algebras will be essentially different from the purely algebraic approach to zLpd and zJpd algebras in Part I. We will first study continuous bilinear functionals φ with the property that $xy = yx = 0$ implies $\varphi(x, y) = 0$ and then apply the obtained results to show that, under certain additional assumptions, zpd Banach algebras are also zLpd and zJpd Banach algebras. This is roughly the concept of the chapter, although we will make several digressions.

6.1 The Condition $xy = yx = 0$

Let A be a Banach algebra. As always, we write

$$[x, y] = xy - yx \quad \text{and} \quad x \circ y = xy + yx$$

for $x, y \in A$. As should be evident from the title, in this chapter we will be mainly concerned with a continuous bilinear functional $\varphi \colon A \times A \to \mathbb{C}$ that satisfies either

$$[x, y] = 0 \implies \varphi(x, y) = 0$$

or

$$x \circ y = 0 \implies \varphi(x, y) = 0.$$

In each of the two cases, φ also satisfies

$$xy = yx = 0 \implies \varphi(x, y) = 0. \tag{6.1}$$

We will therefore first consider (6.1). This will be the sole topic of the present section. The passage to zero Lie/Jordan product determined Banach algebras in subsequent sections will be rather smooth. We remark that such an approach has not turned out so fruitful in the algebraic theory; see the final part of Sect. 3.3.

What result to expect if φ satisfies (6.1)? The most desirable conclusion would be that

$$\varphi(x, y) = \tau_1(xy) + \tau_2(yx)$$

for some $\tau_1, \tau_2 \in A'$. Our first theorem does not yet establish this but makes a crucial step toward it. It should be remarked that Theorem 3.23, which studies the same condition for bilinear functionals on unital algebras, is its weak algebraic version. The concept of proof is the same, but we provide the details anyway.

Theorem 6.1 *Let A be a zpd Banach algebra having a bounded approximate identity. If $\varphi: A \times A \to \mathbb{C}$ is a continuous bilinear functional satisfying $\varphi(x, y) = 0$ whenever $xy = yx = 0$, then*

$$\varphi(xy, zw) + \varphi(wx, yz) = \varphi(x, yzw) + \varphi(wxy, z) \quad (x, y, z, w \in A), \tag{6.2}$$

and there exists a $\sigma \in A'$ such that

$$\varphi(xy, z) - \varphi(y, zx) + \varphi(yz, x) = \sigma(xyz) \quad (x, y, z \in A). \tag{6.3}$$

Moreover, if φ is symmetric, then

$$\varphi(x, y) = \frac{1}{2}\sigma(x \circ y) \quad (x, y \in A). \tag{6.4}$$

Proof Fix, temporarily, $x', y' \in A$ such that $x'y' = 0$. Define a continuous bilinear functional $\varphi': A \times A \to \mathbb{C}$ by

$$\varphi'(x, y) = \varphi(y'x, yx').$$

Note that $xy = 0$ implies $y'x \cdot yx' = yx' \cdot y'x = 0$, and therefore $\varphi'(x, y) = 0$. As A is zpd and hence has property $\mathbb{B}$, it follows that

$$\varphi'(xy, z) = \varphi'(x, yz) \quad (x, y, z \in A);$$

that is,

$$\varphi(y'xy, zx') - \varphi(y'x, yzx') = 0 \quad (x, y, z \in A).$$

Now fix $x, y, z \in A$. Note that the result of the preceding paragraph can be stated as follows: the continuous bilinear functional $\varphi'' : A \times A \to \mathbb{C}$ defined by

$$\varphi''(x', y') = \varphi(y'xy, zx') - \varphi(y'x, yzx')$$

has the property that $\varphi''(x', y') = 0$ whenever $x', y' \in A$ are such that $x'y' = 0$. Consequently, φ'' satisfies

$$\varphi''(x'y', z') = \varphi(x', y'z') \quad (x', y', z' \in A).$$

That is,

$$\varphi(z'xy, zx'y') - \varphi(z'x, yzx'y') = \varphi(y'z'xy, zx') - \varphi(y'z'x, yzx') \tag{6.5}$$

for all $x', y', z' \in A$ as well as for all $x, y, z \in A$. Observe that all terms in (6.5) involve $z'x$ and zx'. Since $A^2 = A$ by Cohen's factorization theorem, (6.5) implies (6.2) (indeed, just write x for $z'x$, z for zx', and w for y').

Take a bounded approximate identity $(e_\lambda)_{\lambda \in \Lambda}$ for A. Replacing z by e_λ in (6.2), we see that the net $(\varphi(wxy, e_\lambda))_{\lambda \in \Lambda}$ is convergent. We can thus define a linear functional σ on A^3 by

$$\sigma(u) = \lim_{\lambda \in \Lambda} \varphi(u, e_\lambda).$$

However, $A^3 = A$, so σ is actually defined on A; moreover, as the net $(e_\lambda)_{\lambda \in \Lambda}$ is bounded, it follows that σ is continuous, so $\sigma \in A'$. Furthermore, by (6.2), we have

$$\lim_{\lambda \in \Lambda} \big(\varphi(xy, e_\lambda w) - \varphi(x, ye_\lambda w) + \varphi(wx, ye_\lambda) - \varphi(wxy, e_\lambda)\big) = 0,$$

which gives

$$\varphi(xy, w) - \varphi(x, yw) + \varphi(wx, y) - \sigma(wxy) = 0 \quad (x, y, w \in A).$$

Observe that is just another form of (6.3).

Finally, assume that φ is symmetric. Substituting e_λ for y in (6.3), we obtain

$$\varphi(xe_\lambda, z) - \varphi(zx, e_\lambda) + \varphi(x, e_\lambda z) = \sigma(xe_\lambda z).$$

By taking limits, we obtain

$$\varphi(x, z) - \sigma(zx) + \varphi(x, z) = \sigma(xz),$$

which yields (6.4).   □

Remark 6.2 Theorem 6.1 holds for continuous bilinear maps φ from $A \times A$ to any normed space X, not only to $\mathbb{C}$. Indeed, one just has to apply Proposition 4.9 (iv) in the proof.

Every bilinear functional $\varphi \colon A \times A \to \mathbb{C}$ can be written as

$$\varphi = \varphi_1 + \varphi_2,$$

where φ_1 is a symmetric and φ_2 is a skew-symmetric bilinear functional. Indeed, we define φ_1 and φ_2 by

$$\varphi_1(x, y) = \frac{1}{2}\big(\varphi(x, y) + \varphi(y, x)\big)$$

and

$$\varphi_2(x, y) = \frac{1}{2}\big(\varphi(x, y) - \varphi(y, x)\big).$$

Obviously, if φ is continuous, then so are φ_1 and φ_2, and if φ satisfies (6.1), then so do φ_1 and φ_2. Theorem 6.1 describes the form of φ_1, so it remains to treat φ_2. That is to say, it is enough to consider the case where φ in Theorem 6.1 is skew-symmetric. The goal is then to show that

$$\varphi(x, y) = \rho([x, y])$$

for some $\rho \in A'$. However, Example 3.28 shows that this is not always true. Indeed, the algebra A from this example is finite-dimensional, and so by choosing $F = \mathbb{C}$, we can view it as a Banach algebra. Some additional assumptions are thus necessary to reach our goal. We need a few more definitions to state them.

Let A be a Banach algebra. Recall that a Banach space M is called a *Banach A-bimodule* if it is an A-bimodule (in the algebraic sense) and there exists a positive constant M such that

$$\|x \cdot m\| \le M\|x\|\|m\| \quad \text{and} \quad \|m \cdot x\| \le M\|m\|\|x\| \quad (m \in M, x \in A).$$

Note that we will always denote module multiplications by $\cdot$.

We are familiar with the following definition in the case where $M = A$.

Definition 6.3 Let A be an algebra, and let M be an A-bimodule. A linear map $\delta : A \to M$ is called a *derivation* if

$$\delta(xy) = \delta(x) \cdot y + x \cdot \delta(y) \quad (x, y \in A).$$

If there exists an $m \in M$ such that $\delta(x) = m \cdot x - x \cdot m$ for all $x \in A$, then δ is a called an *inner derivation*.

Of course, inner derivations indeed are derivations. A simple example of a derivation that is not inner is the map $\delta : C^1[0, 1] \to C[0, 1]$, $\delta(f) = f'$.

Let A be a Banach algebra. The dual space A' of A becomes a Banach A-bimodule by defining

$$(x \cdot f)(y) = f(yx), \quad (f \cdot x)(y) = f(xy) \quad (x, y \in A, f \in A').$$

Definition 6.4 A Banach algebra A is said to be *weakly amenable* if every continuous derivation from A to A' is inner.

For us, the two important examples of weakly amenable Banach algebras are C^*-algebras [76, Theorem 5.6.77] and group algebras $L^1(G)$ for any locally compact group G [76, Theorem 5.6.48]. Recall that these Banach algebras also satisfy the assumptions of Theorem 6.1, i.e., they are zpd and have bounded approximate identities. This speaks in favor of the following two theorems.

Theorem 6.5 *Let A be a weakly amenable zpd Banach algebra having a bounded approximate identity. If $\varphi : A \times A \to \mathbb{C}$ is a continuous skew-symmetric bilinear functional satisfying $\varphi(x, y) = 0$ whenever $xy = yx = 0$, then there exists a $\rho \in A'$ such that*

$$\varphi(x, y) = \rho([x, y]) \quad (x, y \in A).$$

Proof By Theorem 6.1, there is a $\sigma \in A'$ satisfying (6.3). Since φ is skew-symmetric, we can write this identity as

$$\sigma(xyz) = \varphi(xy, z) + \varphi(zx, y) + \varphi(yz, x) \quad (x, y, z \in A). \tag{6.6}$$

This readily implies that

$$\sigma(xyz) = \sigma(yzx) \quad (x, y, z \in A). \tag{6.7}$$

Define $\delta : A \to A'$ by

$$\delta(x)(z) = \sigma(xz) + \varphi(x, z).$$

We claim that δ is a continuous derivation. The linearity and continuity are clear, so we only have to check the derivation law. We have

$$\big(\delta(x) \cdot y + x \cdot \delta(y)\big)(z)$$
$$= \delta(x)(yz) + \delta(y)(zx)$$
$$= \sigma(xyz) + \varphi(x, yz) + \sigma(yzx) + \varphi(y, zx).$$

Using (6.7) and skew-symmetry of φ, it follows that

$$\big(\delta(x) \cdot y + x \cdot \delta(y)\big)(z) = 2\sigma(xyz) - \varphi(yz, x) - \varphi(zx, y).$$

We may now apply (6.6) to conclude that

$$\big(\delta(x) \cdot y + x \cdot \delta(y)\big)(z) = \sigma(xyz) + \varphi(xy, z) = \delta(xy)(z).$$

This proves that δ is a derivation.

Since A is weakly amenable, there exists a $\rho \in A'$ such that

$$\delta(x) = \rho \cdot x - x \cdot \rho \quad (x \in A).$$

Accordingly,

$$\delta(x)(y) = \rho([x, y]) \quad (x, y \in A),$$

and hence, by the definition of δ,

$$\rho([x, y]) - \varphi(x, y) = \sigma(xy) \quad (x, y \in A). \tag{6.8}$$

We can write (6.7) as $\sigma([x, yz]) = 0$. Since $A^2 = A$ by Cohen's factorization theorem, this means that σ vanishes on commutators. Accordingly, the bilinear functional on the right-hand side of (6.8) is symmetric. On the other hand, the bilinear functional on the left-hand side of (6.8) is obviously skew-symmetric. Therefore, both sides are zero. That is, $\sigma = 0$ and $\rho([x, y]) - \varphi(x, y) = 0$ for all $x, y \in A$, which is the desired conclusion. $\quad\square$

Combining Theorems 6.1 and 6.5, we obtain the main result of this section.

Theorem 6.6 *Let A be a weakly amenable zpd Banach algebra having a bounded approximate identity. If $\varphi \colon A \times A \to \mathbb{C}$ is a continuous bilinear functional satisfying $\varphi(x, y) = 0$ whenever $xy = yx = 0$, then there exist $\tau_1, \tau_2 \in A'$ such that*

$$\varphi(x, y) = \tau_1(xy) + \tau_2(yx) \quad (x, y \in A). \tag{6.9}$$

Proof As shown in the discussion following the proof of Theorem 6.1, $\varphi = \varphi_1' + \varphi_2$, where φ_1 is a symmetric continuous bilinear functional, φ_2 is a skew-symmetric continuous bilinear functional, and both φ_1 and φ_2 satisfy the assumption of the theorem. By Theorem 6.1, there is a $\tau \in A'$ such that

$$\varphi_1(x, y) = \tau(x \circ y) \quad (x, y \in A),$$

and by Theorem 6.5, there is a $\rho \in A'$ such that

$$\varphi_2(x, y) = \rho([x, y]) \quad (x, y \in A).$$

Setting $\tau_1 = \tau + \rho$ and $\tau_2 = \tau - \rho$, we thus see that (6.9) holds. $\qquad\square$

Remark 6.7 In the case where A is a C^*-algebra, Theorem 6.6 follows from a theorem by Goldstein [86, Theorem 1.10]. Stated in a slightly different, but equivalent form (see [91, Theorem 3.1]), this theorem states that a continuous bilinear functional $\varphi : A \times A \to \mathbb{C}$ is of the form (6.9) provided that $\varphi(x, y) = 0$ whenever x and y are self-adjoint and satisfy $xy = 0$ (of course, $yx = (xy)^* = 0$ then holds as well).

Theorem 6.1 was obtained, in some form, in [7], while other results of this section are taken from [19]. We also mention the paper [115], which contains an algebraic version of Theorem 6.6 for algebras of matrices over finite-dimensional central division algebras, and also the paper [8], which derives the same conclusion in the case where A is the algebra of matrices over a field, however, under a considerably milder assumption that $\varphi(x, y) = 0$ holds only for pairs of orthogonal rank one idempotents x and y (see also an application of this result in [97]).

6.2   zLpd Banach Algebras

A zLpd Banach algebra is, of course, a Banach algebra in which every continuous bilinear functional $\varphi : A \times A \to \mathbb{C}$ that vanishes on pairs of commuting elements is of the form $\varphi(x, y) = \tau([x, y])$ for some $\tau \in A'$. This can also be stated as follows.

Definition 6.8 A Banach algebra A is *zero Lie product determined*, or *zLpd* for short, if the Lie–Banach algebra A^- is zero product determined.

Example 6.9 A commutative Banach algebra is obviously a zLpd Banach algebra.

Example 6.10 A non-zLpd algebra from Example 3.8 is finite-dimensional, and so it is also a non-zLpd Banach algebra (when taking $\mathbb{C}$ for F).

Our main result on zLpd Banach algebras is a direct consequence of Theorem 6.5.

Theorem 6.11 *If A is a weakly amenable zpd Banach algebra with a bounded approximate identity, then A is a zLpd Banach algebra.*

Proof Let $\varphi: A \times A \to \mathbb{C}$ be a continuous bilinear functional with the property that $[x, y] = 0$ implies $\varphi(x, y) = 0$. Then $\varphi(x, x) = 0$ for all $x \in A$. Replacing x by $x + y$ in this identity, we see that φ is skew-symmetric. Theorem 6.5 therefore yields the desired conclusion.   □

As mentioned before the statement of Theorem 6.5, all C^*-algebras and all group algebras $L^1(G)$ satisfy the conditions of Theorem 6.11. We can thus state the following corollaries.

Corollary 6.12 *Every C^*-algebra is a zLpd Banach algebra.*

Corollary 6.13 *For every locally compact group G, the group algebra $L^1(G)$ is a zLpd Banach algebra.*

In the rest of the section, we present an alternative approach to zLpd Banach algebras (which was also discovered in [19]). Its only drawback is that it does not yield many examples of zLpd Banach algebras that are not already covered by Theorem 6.11.

We need some definitions.

Definition 6.14 Let A be an algebra, and let M be an A-bimodule. A linear map $\delta : A \to M$ is called a *Jordan derivation* if

$$\delta(x^2) = \delta(x) \cdot x + x \cdot \delta(x) \quad (x \in A).$$

Of course, every derivation is a Jordan derivation. The question of when the converse is true has been studied since the 1950s [95] by many authors working in different areas; see [43, 104, 121] where further references can be found.

As in the preceding section, we will be interested in the situation where A is a Banach algebra and $M = A'$.

Definition 6.15 Let A be a Banach algebra. A Jordan derivation $\delta : A \to A'$ is said to be *cyclic* if

$$\delta(x)(y) = -\delta(y)(x) \quad (x, y \in A).$$

Every inner derivation from A to A' is obviously a continuous cyclic Jordan derivation. We will now show that Banach algebras for which the converse statement holds are zLpd.

Theorem 6.16 *Let A be a Banach algebra. If every continuous cyclic Jordan derivation* $\delta : A \to A'$ *is an inner derivation, then A is a zLpd Banach algebra.*

Proof Let $\varphi : A \times A \to \mathbb{C}$ be a continuous bilinear functional satisfying $\varphi(x, y) = 0$ whenever $[x, y] = 0$. In particular,

$$\varphi(x, x) = 0 \quad (x \in A),$$

so φ is skew-symmetric. Furthermore,

$$\varphi(x^2, x) = 0 \quad (x \in A).$$

Substituting $x \pm y$ for x, we arrive at

$$\varphi(x^2, y) + \varphi(x \circ y, x) = 0 \quad (x, y \in A). \tag{6.10}$$

Define $\delta : A \to A'$ by

$$\delta(x)(y) = \varphi(x, y) \quad (x, y \in A).$$

Using (6.10) along with the skew-symmetry of φ, we obtain

$$\begin{aligned}
\delta(x^2)(y) &= \varphi(x^2, y) \\
&= \varphi(x, x \circ y) \\
&= \delta(x)(xy) + \delta(x)(yx) \\
&= \big(\delta(x) \cdot x + x \cdot \delta(x)\big)(y)
\end{aligned}$$

for all $x, y \in A$. This means that δ is a Jordan derivation. Since φ is skew-symmetric, δ is cyclic. Therefore, by our assumption, δ is an inner derivation. Hence, there exists a $\tau \in A'$ such that $\varphi(x, y) = \tau([x, y])$ for all $x, y \in A$.   □

Remark 6.17 Note that we did not use the assumption that φ vanishes on pairs of commuting elements in its full generality, but only that $\varphi(x, x) = \varphi(x, x^2) = 0$ for every $x \in A$. The proof thus yields a somewhat better result than stated in Theorem 6.16.

In order to apply Theorem 6.16 to a concrete situation, two problems emerge:

(1) Is every continuous cyclic Jordan derivation from A to A' a derivation?
(2) Is every continuous cyclic derivation from A to A' inner?

A Banach algebra A for which the question (2) has an affirmative answer is called a *cyclically amenable* Banach algebra. Every weakly amenable Banach algebra is obviously cyclically amenable. For examples of cyclically amenable Banach algebras that are not weakly amenable, see the seminal paper [90] and also [19].

In a cyclically amenable Banach algebra, we still have to deal with the question (1). The following corollary to Theorem 6.16 thus holds.

Corollary 6.18 *Let A be a cyclically amenable Banach algebra. If every continuous cyclic Jordan derivation from A to A' is a derivation, then A is a zLpd Banach algebra.*

In [104], Johnson proved that if A is either a C^*-algebra or a group algebra $L^1(G)$, where G is an *amenable* locally compact group, then every continuous Jordan derivation from A to a Banach A-bimodule M is a derivation. Thus, Corollary 6.12 and the amenable case of Corollary 6.13 also follow from Corollary 6.18. We do not know whether the same can be said for the general case of Corollary 6.13, i.e., we do not know whether every continuous Jordan derivation from $L^1(G)$ to $L^1(G)'$ is a derivation for any locally compact group G.

As an application of Corollary 6.18, it was shown in [19] that if A is a cyclically amenable Banach algebra with a bounded approximate identity, then the matrix algebra $M_n(A)$, with $n \geq 2$, is a zLpd Banach algebra. Examples of such Banach algebras A include unital Banach algebras polynomially generated by a single element, C^*-algebras, and group algebras $L^1(G)$ for any locally compact group G (see [19] for details).

6.3   zJpd Banach Algebras

Before formally defining the notion of a zJpd Banach algebra, we first introduce a more narrow class of Banach algebras for which Theorem 6.1 immediately yields a definitive result. It is an analytic analogue of the class of symmetrically zJpd algebras (see Definition 3.21).

Definition 6.19 A Banach algebra A is *symmetrically zero Jordan product determined*, or *symmetrically zJpd* for short, if, for every continuous symmetric bilinear functional $\varphi: A \times A \to \mathbb{C}$ satisfying $\varphi(x, y) = 0$ whenever $x, y \in A$ are such that $x \circ y = 0$, there exists a continuous linear functional τ on A such that $\varphi(x, y) = \tau(x \circ y)$ for all $x, y \in A$.

Of course, $xy = yx = 0$ implies $x \circ y = 0$. We thus have the following corollary to Theorem 6.1, which may be viewed as an analytic version of Corollary 3.25.

Corollary 6.20 *Let A be a Banach algebra having a bounded approximate identity. If A is zpd, then A is also symmetrically zJpd.*

Remark 6.21 If A is a commutative Banach algebra, then instead of saying that it is symmetrically zJpd it seems more appropriate to simply say that it is *symmetrically zpd*, i.e., *symmetrically zero product determined* (compare Definition 1.12). Of course, a commutative zpd Banach algebra is also symmetrically zpd. By [9, Corollary 3.1], the algebra of Lipschitz functions $\mathrm{lip}_\alpha(K)$, with $0 < \alpha < 1$ and K a compact metric space, is a commutative symmetrically zpd Banach algebra, and at present it is unknown whether or not it is zpd for every $0 < \alpha < 1$.

Now we proceed to the main theme.

Definition 6.22   A Banach algebra A is *zero Jordan product determined*, or *zJpd* for short, if the Jordan–Banach algebra A^+ is zero product determined.

If written Definition 6.22 in an explicit way in terms of the functional φ, the only difference to Definition 6.19 is that φ is not assumed to be symmetric.

The results from Sect. 6.1 are not as easily applicable to zJpd Banach algebras as they are to zLpd Banach algebras. The reason for this difference is that the existence of elements $x, y \in A$ such that $x \circ y = 0$ but $xy \neq 0$ is not obvious, while we always have an abundance of pairs of commuting elements (in particular, every element commutes with itself). In order to assert that the Banach algebra A is zJpd, we are therefore forced to add an additional assumption to those from Theorems 6.5 and 6.6.

For any algebra A, we write

$$[A, A] = \mathrm{span}\,\{[x, y] \mid x, y \in A\}$$

and

$$[[A, A], [A, A]] = \mathrm{span}\,\{[[x, y], [z, w]] \mid x, y, z, w \in A\}.$$

Of course,

$$[[A, A], [A, A]] \subseteq [A, A],$$

and hence

$$\overline{[[A, A], [A, A]]} \subseteq \overline{[A, A]}$$

if A is a Banach algebra. This inclusion may be proper, but often it is not.

Theorem 6.23 *Let A be a weakly amenable zpd Banach algebra with a bounded approximate identity. If*

$$\overline{[[A, A], [A, A]]} = \overline{[A, A]}, \tag{6.11}$$

then A is a zJpd Banach algebra.

Proof Let $\varphi \colon A \times A \to \mathbb{C}$ be a continuous bilinear functional such that $x \circ y = 0$ implies $\varphi(x, y) = 0$. Assume first that φ is skew-symmetric. Our goal is then to show that $\varphi = 0$.

Theorem 6.5 tells us that there is a $\rho \in A'$ such that

$$\varphi(x, y) = \rho([x, y]) \quad (x, y \in A).$$

Set

$$U = \{u \in A \mid u^2 = 0\}.$$

Observe that

$$(u - v) \circ (u + v) = (u - v)(u + v) + (u + v)(u - v)$$
$$= 2u^2 - 2v^2$$
$$= 0$$

for all $u, v \in U$, which gives

$$\rho([u, v]) = \frac{1}{2}\rho([u - v, u + v]) = 0.$$

This clearly implies that

$$\rho([r, s]) = 0 \quad (r, s \in \overline{\text{span}}\, U). \tag{6.12}$$

Now consider the map $\Phi \colon A \times A \to A/\overline{\text{span}}\, U$ defined by

$$\Phi(x, y) = yx + \overline{\text{span}}\, U$$

(compare Example 1.11). Since $xy = 0$ implies $yx \in U$, Φ is a continuous bilinear map satisfying $\Phi(x, y) = 0$ whenever $xy = 0$. As A is zpd, we have (by Proposition 4.9 (iv))

$\Phi(x, y) = T(xy)$ for some linear map $T : A^2 \to A/\overline{\text{span}}\, U$. In particular, $\Phi(xy, z) = \Phi(x, yz)$ for all $x, y, z \in A$, which means that

$$[zx, y] = z \cdot xy - yz \cdot x \in \overline{\text{span}}\, U$$

for all $x, y, z \in A$. Since $A^2 = A$ by Cohen's factorization theorem, we thus have

$$[A, A] \subseteq \overline{\text{span}}\, U,$$

and so (6.12) gives

$$\rho\big([[A, A], [A, A]]\big) = \{0\}.$$

As ρ is continuous and (6.11) holds, this implies that $\rho([A, A]) = \{0\}$; that is, $\varphi = 0$.

We have thus handled the case where φ is skew-symmetric. Let now φ be arbitrary. Then the continuous bilinear functional

$$(x, y) \mapsto \varphi(x, y) - \varphi(y, x)$$

is skew-symmetric and also vanishes on pairs of anticommuting elements. By what we have just shown, it is equal to zero, which means that φ is symmetric. The desired conclusion that there is a $\tau \in A'$ such that $\varphi(x, y) = \tau(x \circ y)$ for all $x, y \in A$ now follows from Theorem 6.1.   □

Remark 6.24 The fact noticed in the proof that $[A, A] \subseteq \overline{\text{span}}\, U$ is interesting in its own right and will be considered again in Sect. 9.1.

The question that now arises is whether our main examples of weakly amenable zpd Banach algebras with bounded approximate identities, i.e., C^*-algebras and group algebras $L^1(G)$, also satisfy (6.11). This is discussed in [20]. We will not give details here, since this question is slightly out of the scope of the book. Let us only sketch the main ideas.

Every C^*-algebra A indeed satisfies (6.11). By employing the second dual of A, the proof can be reduced to the case where A is a von Neumann algebra. The proof for this case in [20] uses some results obtained in that paper, but there are other ways. In fact, every von Neumann algebra A satisfies

$$A = Z(A) + [A, A],$$

where $Z(A)$ is the center of A. As noticed in the proof of [57, Theorem 5.19], this follows by combining some old results on von Neumann algebras. Of course, $A = Z + [A, A]$ immediately implies $\big[[A, A], [A, A]\big] = [A, A]$.

We thus have the following corollary to Theorem 6.23.

Corollary 6.25 *Every C*-algebra is a zJpd Banach algebra.*

The approach to group algebras $L^1(G)$ is based on amenability and related notions. A Banach algebra A is said to be *amenable* if every continuous derivation from A to M' is inner for any Banach A-bimodule M. An equivalent condition is that A has an *approximate diagonal*. This is a bounded net $(\mathbf{t}_\lambda)_{\lambda \in \Lambda}$ in the projective tensor product $A \widehat{\otimes} A$ such that

$$\lim_{\lambda \in \Lambda} (x \cdot \mathbf{t}_\lambda - \mathbf{t}_\lambda \cdot x) = 0$$

and

$$\lim_{\lambda \in \Lambda} \pi(\mathbf{t}_\lambda)x = x$$

for all $x \in A$, where

$$x \cdot y \otimes z = xy \otimes z, \quad y \otimes z \cdot x = y \otimes zx,$$

and

$$\pi(y \otimes z) = yz$$

for all $x, y, z \in A$.

Let $\mathbf{t} \mapsto \mathbf{t}^\circ$ be the flip map (defined by $(x \otimes y)^\circ = y \otimes x$). A tensor $\mathbf{t} \in A \widehat{\otimes} A$ is said to be *symmetric* if $\mathbf{t}^\circ = \mathbf{t}$. An amenable Banach algebra A that has an approximate diagonal consisting of symmetric tensors is said to be *symmetrically amenable* [104]. Not every amenable Banach algebra is symmetrically amenable, but many are.

Every symmetrically amenable Banach algebra A satisfies the desired condition (6.11). The sketch of proof is as follows. One first checks that for all $x, y \in A$ and $\mathbf{t} \in A \widehat{\otimes} A$,

$$\pi(\mathbf{t})xy - y\pi(\mathbf{t})x + \pi\big((y \cdot \mathbf{t}^\circ - \mathbf{t}^\circ \cdot y) \star x\big) \in \overline{\big[y, [A, A]\big]},$$

where the operation $\star$ is defined by

$$(u \otimes v) \star x = (ux) \otimes v \quad (u, v, x \in A).$$

Taking an approximate diagonal $(\mathbf{t}_\lambda)_{\lambda \in \Lambda}$ consisting of symmetric tensors and writing $\mathbf{t}_\lambda$ for $\mathbf{t}$ in the above relation, one shows that

$$[x, y] \in \overline{\big[y, [A, A]\big]} \quad (x, y \in A).$$

The desired conclusion that

$$[x, y] \in \overline{[[A, A], [A, A]]} \quad (x, y \in A)$$

then easily follows.

One way of defining an amenable group is that it is a locally compact group, G, such that the group algebra $L^1(G)$ is amenable. It turns out that $L^1(G)$ is then also symmetrically amenable. The following corollary to Theorem 6.23 therefore holds.

Corollary 6.26 *For every amenable locally compact group G, the group algebra $L^1(G)$ is a zJpd Banach algebra.*

The question whether this corollary holds for all locally compact groups G is open at present.

We know from Part I that unital algebras (over fields of characteristic different from 2) generated by idempotents are zJpd (Theorem 3.15). From Remark 3.16, it is evident that the same proof also covers the Banach algebra case. Let us record this result.

Theorem 6.27 *If a unital Banach algebra A is generated by idempotents, then A is zJpd.*

Example 6.28 For any unital Banach algebra B and $n \geq 2$, the matrix algebra $M_n(B)$ is zJpd (in light of Corollary 2.4). See also Example 5.15 for further examples.

Finally, we list a few non-examples.

Example 6.29 A non-zpd commutative Banach algebra (such as $C^n[0, 1]$) is also non-zJpd.

Example 6.30 In Example 3.20, we saw that the (complex) Grassmann algebra G_n is a non-zJpd algebra. As it is finite-dimensional, it is also a non-zJpd Banach algebra.

Example 6.31 Example 5.26 shows that there exists a Banach space X such that $\mathscr{B}(X)$ is a non-zpd Banach algebra. Essentially the same proof, based on Theorem 4.13, shows that it is also a non-zJpd Banach algebra.

Part III
Applications

Homomorphisms and Related Maps

7

It is fairly obvious that the zpd condition can be useful in the consideration of linear maps between associative algebras that preserve either zero products or commutativity. Moreover, its introduction was partially motivated by these two topics. It is therefore natural to discuss them in the first two sections of this final part of the book, which is devoted to demonstrating the usefulness of the zpd concept. In the last and third section of this chapter, we will consider the problem of describing Jordan homomorphisms in zpd algebras. The common feature of all the three topics of this chapter is that they involve homomorphism-like maps.

Unlike in Parts I and II, in Part III we will often omit detailed proofs and instead provide references for the complete arguments. Our goal is just to give a clear indication of the applicability of zpd (Banach) algebras.

7.1 Zero Product Preserving Maps

The maps appearing in the title were briefly touched in Example 1.9. We will now discuss them in greater detail. Let us start with the definition.

Definition 7.1 A linear map T from an algebra A to an algebra B is said to be *zero product preserving* if, for all $x, y \in A$,

$$xy = 0 \implies T(x)T(y) = 0.$$

Throughout, it goes without saying that the algebras A and B are associative.

© The Author(s), under exclusive license to Springer Nature Switzerland AG 2021
M. Brešar, *Zero Product Determined Algebras*, Frontiers in Mathematics,
https://doi.org/10.1007/978-3-030-80242-4_7

The study of zero product preserving maps has a long history. We can start by mentioning that in his classical book [28], Banach studied the form of isometries of $L^p([0, 1])$ and made the crucial observation that they send functions with disjoint support into functions with disjoint support. His work was later completed by Lamperti [117]. We remark that zero product preserving maps are also called Lamperti operators or separating maps in some contexts; and indeed there are various contexts, both analytic and algebraic, in which these maps have been considered. A partial list of references includes [25, 26, 30, 67, 68, 119, 124, 134, 158, 168, 169].

The most obvious example of a zero product preserving linear map is an algebra homomorphism, possibly multiplied by an element commuting with any element from its image. There are, however, other examples, like any linear map with the property that the product of any two elements in its image is 0. Under the assumption that the range of a zero product preserving linear map is sufficiently large, this latter example becomes irrelevant and one may wish to prove that the map is close to the one from the first basic example. It is rather obvious that zpd algebras provide a natural framework for considering this problem.

We first record a simple result that only slightly extends the observation from Example 1.9.

Theorem 7.2 *Let A and B be algebras, and let $T : A \to B$ be a zero product preserving linear map. If A is zpd, then there exists a linear map $S : A \to B$ such that*

$$S(xy) = T(x)T(y) \quad (x, y \in A).$$

Moreover, if both A and B are unital and the range of T contains an invertible element of B (in particular, if T is surjective), then $c = T(1)$ is invertible in B and there exists a homomorphism $\Phi : A \to B$ such that

$$T(x) = c\Phi(x) = \Phi(x)c \quad (x \in A).$$

Proof The existence of S of course follows by applying the alternative definition of a zpd algebra given in Proposition 1.3 (iv) to the bilinear map

$$(x, y) \mapsto T(x)T(y).$$

Assume now that A and B are unital and that there exists an $a \in A$ such that $T(a)$ is invertible in B. From $S(xy) = T(x)T(y)$, we see that

$$T(x)c = S(x) = cT(x) \quad (x \in A),$$

and from

$$T(a)^2 = S(a^2) = S(1 \cdot a^2) = cT(a^2),$$

we see that c is invertible. Set

$$\Phi(x) = c^{-1}T(x),$$

and observe that

$$\begin{aligned}
\Phi(xy) &= c^{-1}T(xy) \\
&= c^{-2}T(1)T(xy) \\
&= c^{-2}S(xy) \\
&= c^{-2}T(x)T(y) \\
&= \Phi(x)\Phi(y),
\end{aligned}$$

so Φ is a homomorphism. Obviously, $cT(x) = T(x)c$ implies $c\Phi(x) = \Phi(x)c$.   □

The following analytic version of Theorem 7.2 can be proved in essentially the same way.

Theorem 7.3 *Let A and B be Banach algebras, and let $T : A \rightarrow B$ be a continuous zero product preserving linear map. If A is zpd, then there exists a continuous linear map $S : A^2 \rightarrow B$ such that*

$$S(xy) = T(x)T(y) \quad (x, y \in A).$$

Moreover, if both A and B are unital and the range of T contains an invertible element of B (in particular, if T is surjective), then $c = T(1)$ is invertible in B and there exists a continuous homomorphism $\Phi : A \rightarrow B$ such that

$$T(x) = c\Phi(x) = \Phi(x)c \quad (x \in A).$$

We have recorded these two simple theorems just to give the general idea. Our next result is deeper. To state it, we need some definitions.

Let A be an algebra. A linear map $W : A \rightarrow A$ is called a *centralizer* if

$$W(xy) = W(x)y = xW(y) \quad (x, y \in A).$$

Under standard operations, the set of all centralizers of A is an algebra. We call it the *centroid* of A and denote it by $\Gamma(A)$. For example, if c lies in the center of A, then $x \mapsto cx$ is a centralizer. Observe that every centralizer W is of such a form if A is unital (with $c = W(1)$). However, we are now primarily interested in Banach algebras without unity. So, assume until further notice that A is a Banach algebra. We say that A is *faithful* if, for any $a \in A$, $aA = Aa = \{0\}$ implies $a = 0$. It is easy to see that in such a Banach algebra every centralizer is continuous, and, moreover, $\Gamma(A)$ is a closed unital subalgebra of $\mathscr{B}(A)$, the Banach algebra of all continuous linear operators of A. We also remark that $W \mapsto (W, W)$ defines an isomorphism from $\Gamma(A)$ onto the center of the multiplier algebra $\mathscr{M}(A)$ of A.

A map T from an algebra A to an algebra B is a called a *weighted homomorphism* if there exist an invertible centralizer $W \in \Gamma(B)$ and an algebra homomorphism $\Phi : A \to B$ such that

$$T = W\Phi.$$

Note that weighted homomorphisms preserve zero products.

We can now state [6, Corollary 3.9].

Theorem 7.4 *Let A be a zpd Banach algebra with a bounded left approximate identity, and let B be a faithful Banach algebra satisfying $B^2 = B$. Then every surjective continuous zero product preserving linear map $T : A \to B$ is a weighted homomorphism.*

Let us point out that this theorem is, in particular, applicable to C^*-algebras and group algebras $L^1(G)$. The main idea of its proof is similar to that of the preceding two results, but the actual proof is technically much more complicated. Theorem 7.4 is in fact just a simplified version of more general results [6, Theorems 3.6 and 3.8]. See also [9, 10, 15] for further generalizations and variations.

We return to the algebraic framework. The obvious problem we face is to get rid of the assumption in Theorem 7.2 that the range of T contains an invertible element. This is not easy and it seems necessary to restrict our attention to more concrete algebras. Let us state [54, Theorem 5.2].

Theorem 7.5 *Let D be a division algebra, and let $n \geq 2$. If $T : M_n(D) \to M_n(D)$ is a zero product preserving linear map, then either*

(a) $T(x)T(y) = 0$ *for all $x, y \in M_n(D)$ or*
(b) *there exists an endomorphism Φ of $M_n(D)$ such that*

$$T(x) = c\Phi(x) = \Phi(x)c \quad (x \in M_n(D)),$$

where $c = T(1)$.

We remark that c does not necessarily lie in the center of $M_n(D)$ [54, Example 5.1]. We also remark that [54, Theorem 5.2] actually considers rings rather than algebras, but we stated Theorem 7.5 in the form adjusted to our general framework.

Let us sketch the idea of proof. Since the algebra $M_n(D)$ is zpd (Corollary 2.17), there exists a linear map $S : M_n(D) \to M_n(D)$ satisfying $S(xy) = T(x)T(y)$ for all $x, y \in A$. If c is not invertible, then it has nontrivial kernel from which it is possible to derive that S must be 0; thus, (a) holds in this case. If, however, c is invertible, then we see, as in the proof of Theorem 7.2, that (b) holds.

A similar result, which involves two linear maps S and T satisfying $S(x)T(y) = 0$ whenever $xy = 0$ but, on the other hand, holds only for finite-dimensional central simple algebras that are not division algebras, is established in [46].

In our next theorem, we will see that the applications of our theory are not limited strictly to zpd algebras. Specifically, we will see that it may be enough that the algebra under consideration contains a subalgebra that is zpd.

We start with a lemma, which is indeed almost trivial but paves the way for the type of applications we wish to present. More precisely, we will use it in the proofs of Theorems 7.8, 8.3, and 8.14.

Lemma 7.6 *Let A be a unital algebra, let X be a vector space, and let $\varphi : A \times A \to X$ be a bilinear map such that $\varphi(x, y) = 0$ whenever $xy = 0$. If a unital subalgebra R of A is a zpd algebra, then*

$$\varphi(xr, y) = \varphi(x, ry) \quad (x, y \in A, r \in R).$$

Proof Fix $x, y \in A$, and define $\overline{\varphi} : R \times R \to X$ by

$$\overline{\varphi}(r, s) = \varphi(xr, sy).$$

Note that $rs = 0$ implies $xr \cdot ys = 0$, and hence $\overline{\varphi}(r, s) = \varphi(xr, sy) = 0$. By Proposition 1.3 (iv), $\overline{\varphi}$ in particular satisfies $\overline{\varphi}(r, 1) = \overline{\varphi}(1, r)$ for every $r \in R$, which is exactly what we wish to prove. $\qquad\square$

The conditions considered in the next lemma may seem very special, but they will occur in the proofs of Theorems 7.8 and 8.3.

Lemma 7.7 *Let A be a unital algebra, let X be a vector space, and let $\Psi : A^4 \to X$ be a 4-linear map such that for every $i = 1, 2, 3$,*

$$x_i x_{i+1} = 0 \implies \Psi(x_1, x_2, x_3, x_4) = 0.$$

Suppose that a unital subalgebra R of A is a zpd algebra and that a right ideal I of A is contained in R. Then

$$\Psi(u, x, y, z) = \Psi(u, x, 1, yz) \quad (x, y, z \in A, u \in I).$$

Proof By fixing x_3 and x_4, we see from Lemma 7.6 that

$$\Psi(r, y, x_3, x_4) = \Psi(1, ry, x_3, x_4)$$

for all $r \in R$ and $y \in A$. Similarly,

$$\Psi(x_1, r, y, x_4) = \Psi(x_1, 1, ry, x_4)$$

and

$$\Psi(x_1, x_2, r, y) = \Psi(x_1, x_2, 1, ry).$$

Take $u \in I$. Then $uA \subseteq R$, and so it follows from the above identities that

$$\begin{aligned}
\Psi(u, x, y, z) &= \Psi(1, ux, y, z) \\
&= \Psi(1, 1, uxy, z) \\
&= \Psi(1, 1, 1, uxyz) \\
&= \Psi(1, 1, ux, yz) \\
&= \Psi(1, ux, 1, yz) \\
&= \Psi(u, x, 1, yz)
\end{aligned}$$

for all $x, y, z \in A$.   □

Recall that an algebra B is prime if the product of two nonzero ideals in B is always nonzero. It is well-known, and easy to see, that this is equivalent to the condition that for any $a, b \in B$, $aBb = \{0\}$ implies $a = 0$ or $b = 0$. The class of prime algebras is fairly large. In particular, it includes the classes of simple and primitive algebras.

The proof of the next theorem is based on the fact that unital algebras generated by idempotents are zpd. However, all we will require is that our algebra contains only one noncentral idempotent (i.e., an idempotent not contained in the center).

Theorem 7.8 *Let A be a unital algebra containing a noncentral idempotent e, and let B be a prime unital algebra. If T : A → B is a bijective zero product preserving linear map, then c = T(1) is an invertible element lying in the center of B, and there exists an* *isomorphism* $\Phi : A \to B$ *such that*

$$T(x) = c\Phi(x) \quad (x \in A).$$

Proof Let R be the subalgebra of A generated by all idempotents in A. By Theorem 2.15, R is a zpd algebra. Furthermore, Lemma 2.26 tells us that the ideal of A generated by all commutators $|e, x], x \in A$, is contained in R.

Define $\Psi : A^4 \to B$ by

$$\Psi(x_1, x_2, x_3, x_4) = T(x_1)T(x_2)T(x_3)T(x_4).$$

As T preserves zero products, $\Psi(x_1, x_2, x_3, x_4) = 0$ whenever $x_i x_{i+1} = 0$ for some $i = 1, 2, 3$. Thus, the conditions of Lemma 7.7 are met, and therefore

$$\Psi(u, x, y, z) = \Psi(u, x, 1, yz) \quad (u \in I, x, y, z \in A);$$

that is,

$$T(u)T(x)\big(T(y)T(z) - cT(yz)\big) = 0 \quad (x, y, z \in A, u \in I).$$

Since $T(u) \neq 0$ for any nonzero $u \in A$, and since T is surjective and B is prime, it follows that

$$T(y)T(z) = cT(yz) \quad (y, z \in A).$$

Setting $z = 1$, we see that c commutes with every $T(y)$, so it lies in the center of B. Let $a \in A$ be such that $T(a) = 1$. Then

$$1 = 1 \cdot 1 = T(a)T(a) = cT(a^2),$$

proving that c is invertible. It is now straightforward to check that $\Phi : A \to B$ defined by

$$\Phi(x) = c^{-1}T(x)$$

is an isomorphism. $\square$

Example 7.9 The algebra $\mathcal{B}(X)$ of all continuous linear operators on a Banach space X is not always zpd (not even as a Banach algebra, see Example 5.26) but has plenty of noncentral idempotents. Moreover, it is prime. Thus, Theorems 7.2, 7.3, and 7.4 are not applicable to zero product preserving linear maps from $\mathcal{B}(X)$ to $\mathcal{B}(Y)$, but Theorem 7.8 is.

Theorem 7.8 is taken from [44]. This paper was actually written before the notion of a zpd algebra was introduced. The proof we just gave is more conceptual than the one in [44], but the main ideas are the same. Let us also mention that one can establish a more general version of Theorem 7.8, which holds in algebras without unity, but the statement is more technical and, in particular, involves the notion of the extended centroid; see [44, Corollary 4.3] and also [68, Theorem 1].

In the remaining of the section, we briefly discuss a certain variation of the notion of a zero product preserving map. There are actually many variations that appear in the literature. For example, the papers [69, 70, 98, 99, 123] consider the condition that T preserves zero Jordan products, that is, for all $x, y \in A$, $x \circ y = 0$ implies $T(x) \circ T(y) = 0$ (here, as always, $x \circ y$ stands for $xy + yx$). The next definition presents the most obvious example of such a map.

Definition 7.10 A linear map Φ from an algebra A to an algebra B satisfying

$$\Phi(x \circ y) = \Phi(x) \circ \Phi(y) \quad (x, y \in A)$$

is called a *Jordan homomorphism*.

For now, we give only the definition. The problem of describing the form of Jordan homomorphisms will be considered in Sect. 7.3.

Under the assumption that a zero Jordan product preserving linear map is surjective, one usually wants to prove that it is close to a Jordan homomorphism. In a zpd algebra A, we may tackle a somewhat more general problem on maps satisfying the condition that

$$xy = yx = 0 \implies T(x) \circ T(y) = 0.$$

Indeed, Theorem 3.23 (together with Remark 3.24) shows that if A is unital, then the bilinear map

$$\Psi(x, y) = T(x) \circ T(y)$$

satisfies

$$2\Psi(x, y) = \Psi(x \circ y, 1) \quad (x, y \in A),$$

that is,

$$2T(x) \circ T(y) = T(x \circ y) \circ T(1) \quad (x, y \in A).$$

So, for example, if $T(1) = 1$ and the characteristic of the underlying field is not 2, then T is a Jordan homomorphism. Without assuming that T sends 1 to 1, the problem becomes challenging. However, we will not elaborate on this here.

In zpd Banach algebras, we can use Theorem 6.1 (together with Remark 6.2) instead of Theorem 3.23. Obviously, T should now be a continuous linear map from a zpd Banach algebra A to a Banach algebra B. However, A now does not need to have a unity but only a bounded approximate identity. A natural conclusion in this context is that T is a *weighted Jordan homomorphism*. By this, we of course mean that $T = W\Phi$, where $W \in \Gamma(B)$ is an invertible centralizer and $\Phi : A \to B$ is a Jordan homomorphism.

The following result was proved in [7] as an application of (a version of) Theorem 6.1.

Theorem 7.11 *Let A and B be C^*-algebras, and let $T: A \to B$ be a continuous surjective linear map with the property that $T(x) \circ T(y) = 0$ whenever $xy = yx = 0$. Then T is a weighted Jordan homomorphism.*

7.2   Commutativity Preserving Maps

Again, let A and B be associative algebras. We continue to consider zero product preserving linear maps from A to B, but this time with respect to the Lie product $[x, y] = xy - yx$. That is, we will be interested in maps from the following definition.

Definition 7.12 A linear map $T : A \to B$ is said to be *commutativity preserving* if, for all $x, y \in A$,

$$xy = yx \implies T(x)T(y) = T(y)T(x).$$

The problem of describing such maps is, just as the one studied in the preceding section, one of the most studied linear preserver problems (this is a common name for problems of characterizing linear maps that preserve certain properties). Its roots are in linear algebra [167], but through the years it has spread across different areas of abstract algebra and functional analysis. We refer to [56, pp. 218–219] for a historical survey. There has been further development after the publication of this book, but mostly concerning certain generalizations, like dropping the linearity assumption. Our aim is to discuss the usefulness of the zpd notion in the study of the classical linear commutativity preservers.

Besides homomorphisms, there are also *antihomomorphisms*, i.e., linear maps $\Phi : A \to B$ satisfying

$$\Phi(xy) = \Phi(y)\Phi(x) \quad (x, y \in A),$$

which obviously preserve commutativity. Furthermore, if $\Phi : A \to B$ is a homomorphism or an antihomomorphism, c an element of the center $Z(B)$ of B, and $\mu : A \to Z(B)$ a linear map, then $T : A \to B$ defined by

$$T(x) = c\Phi(x) + \mu(x) \tag{7.1}$$

also preserves commutativity. Another example of a commutativity preserver is any linear map from A to B whose image consists of elements commuting with each other.

It turns out that the examples from the preceding paragraph are in fact the only examples of linear commutativity preservers in the case where $A = B$ is the matrix algebra $M_n(F)$, with $n \neq 2$ (incidentally, Φ from (7.1) then takes one of the forms $\Phi(x) = sxs^{-1}$ or $\Phi(x) = sx^t s^{-1}$, where s is an invertible matrix and x^t stands for the transpose of x). This was proved in the 2006 paper [53]. The first step of the proof was showing that, using the present terminology, the algebra $A = M_n(F)$ is zLpd (which is a special case of Corollary 3.11). This implies that for any commutativity preserving linear map $T : A \to A$, there exists a linear map $S : A \to A$ such that

$$S([x, y]) = [T(x), T(y)] \quad (x, y \in A). \tag{7.2}$$

The second step of the proof was finding the connection between S and T. This was incomparably more involved than the technically simple first step. However, this first step was absolutely essential in making a breakthrough in this problem, which was at the time known among experts. In retrospect, [53], being along with [4] the first paper in which a version of the zpd condition was considered, has played an important role in the development of the theory presented in this book.

Before stating the result of [53] precisely and in full generality, we make two remarks. The first one is that the assumption that $n \neq 2$ is necessary.

Example 7.13 It is easy to see that two matrices x and y in $M_2(F)$ commute if and only if x, y, and 1 are linearly dependent. Hence, every linear map that sends 1 to a scalar multiple of 1 preserves commutativity. It is therefore easy to find examples different from those described above.

The second remark is that one does not need to assume that T sends all commuting pairs to commuting pairs, but only that $T(x)$ always commutes with $T(x^2)$; as a matter of fact, any bilinear map φ from $M_n(F) \times M_n(F)$ to some vector space that satisfies

$$\varphi(x, x) = \varphi(x, x^2) = 0$$

for every matrix x is necessarily of the form $\varphi(x, y) = S([x, y])$ [53, Theorem 2.1]. This makes it possible to use the scalar extension argument to establish the result not only for $M_n(F)$, $n \neq 2$, but also for any finite-dimensional *central simple algebra* of dimension different from 4. These are algebras isomorphic to $M_n(D)$, where $n \geq 1$ and D is a finite-dimensional division algebra (of dimension different from 4 if $n = 1$ and of dimension different from 1 if $n = 2$) whose center consists of scalar multiples of 1.

The following is [53, Theorem 4.11]. Note the similarity with Theorem 7.5.

Theorem 7.14 *Let A be a finite-dimensional central simple algebra over a field F such that $\dim_F A \neq 4$. If a linear map $T : A \to A$ satisfies $[T(x^2), T(x)] = 0$ for all $x \in A$ (in particular, if T preserves commutativity), then either*

(a) *$[T(x), T(y)] = 0$ for all $x, y \in A$ or*
(b) *there exist a $c \in F$, an endomorphism or an antiendomorphism Φ of A, and a linear functional μ on A such that*

$$T(x) = c\Phi(x) + \mu(x)1 \quad (x \in A).$$

In more complex algebras, one can combine maps of types (a) and (b) to produce much more involved examples, and so giving a precise description of an arbitrary commutativity preserving linear map may be next to impossible. The assumption one usually makes is that T is surjective or bijective. Then the theory of *functional identities* is applicable, which we briefly and informally recapitulate in the next lines.

Roughly speaking, a functional identity on a ring A is an identical relation in A, which involves maps that are considered as unknowns. An example, which is both important and typical, is the identity

$$\sum_{i=1}^{d} E_i(\overline{x}_d^i)x_i + \sum_{j=1}^{d} x_j F_j(\overline{x}_d^j) = 0 \quad (x_1, \ldots, x_d \in A), \tag{7.3}$$

where $d \geq 2$,

$$\overline{x}_d^i = (x_1, \ldots, x_{i-1}, x_{i+1}, \ldots, x_d) \in A^{d-1}$$

and

$$E_i, F_j : A^{d-1} \to A$$

are the unknown maps. The goal is to describe them. A natural possibility is that they are of the form

$$E_i(\overline{x}_d^i) = \sum_{j \neq i} x_j p_{ij}(\overline{x}_d^{ij}) + \lambda_i(\overline{x}_d^i),$$

$$F_j(\overline{x}_d^j) = -\sum_{i \neq j} p_{ij}(\overline{x}_d^{ij})x_i - \lambda_j(\overline{x}_d^j),$$

where

$$\overline{x}_d^{ij} = \overline{x}_d^{ji} = (x_1, \ldots, x_{i-1}, x_{i+1}, \ldots, x_{j-1}, x_{j+1} \ldots, x_d) \in A^{d-2}$$

and

$$p_{ij} : A^{d-2} \to A, \quad \lambda_i : A^{d-1} \to Z(A)$$

are any maps; here, $Z(A)$ stands for the center of A. If this is the only possibility, i.e., any maps E_i, F_j satisfying (7.3) are of this form, then we say that the functional identity (7.3) has only *standard solutions* on A. The following example is illustrative: if $A = M_n(F)$, then (7.3) has only standard solutions if and only if $n \geq d$. This statement can be appropriately extended to various general classes of rings. In particular, there are many rings in which there are no other solutions of (7.3) than the standard ones. A ring having this, along with some related properties, is said to be *d-free*. This notion is of crucial importance in the theory of functional identities, but we omit the precise definition. The interested reader is referred to [56].

It is easy to explain why functional identities are applicable to the problem of describing a bijective linear commutativity preserver $T : A \to B$. The key point is that $T(x^2)$ commutes with $T(x)$ for every $x \in A$. Note that this can be written as

$$[T(T^{-1}(y)^2), y] = 0$$

for every $y \in B$. Using the standard process of linearization (based on replacing an element by the sum of two elements), we infer from this identity that

$$B(x, y) = T(T^{-1}(x)T^{-1}(y)) \tag{7.4}$$

satisfies

$$[B(x, y) + B(y, x), z] + [B(z, x) + B(x, z), y] + [B(y, z) + B(z, y), x] = 0 \tag{7.5}$$

for all $x, y, z \in B$. Observe that (7.5) is a functional identity of the form (7.3), with

$$E_i(x, y) = -F_j(x, y)$$
$$= B(x, y) + B(y, x)$$
$$= T(T^{-1}(x) \circ T^{-1}(y)) \quad (i, j = 1, 2, 3)$$

(here, as always, $\circ$ denotes the Jordan product). Assuming that B is 3-free, it follows that these maps are necessarily standard solutions of (7.5). This gives information on the action of T on the Jordan product $a \circ b$ of any two elements $a, b \in A$, from which it can be derived, under some additional assumptions, that T is of the form (7.1) with c invertible and Φ an isomorphism, an antiisomorphism, or, sometimes, just a Jordan isomorphism. We refer to [56, Section 7.1] for details. Let us only add an example that in particular shows that some additional assumptions are really necessary.

Example 7.15 Let A be an algebra with an infinite-dimensional center $Z(A)$. Take an injective but not surjective linear map $\gamma : Z(A) \to Z(A)$. Next, choose a subspace Z_1 of $Z(A)$ and a subspace V of A such that

$$Z(A) = \gamma(Z(A)) \oplus Z_1 \quad \text{and} \quad A = V \oplus Z(A).$$

Let δ be a bijective linear map from some algebra A' onto Z_1. Define $T : A \times A' \to A$ by

$$T(v + z, x) = v + \gamma(z) + \delta(x) \quad (v \in V, z \in Z(A), x \in A').$$

Observe that T is a bijective commutativity preserving linear map which maps the ideal $J = \{0\} \times A'$ to the center of A. Therefore, T^{-1} does not preserve commutativity if A' is not commutative.

This example is taken from [49]; another example with a similar message is [56, Example 7.2]. The point we wish to make here is that T is not of the form (7.1) with c invertible and Φ an isomorphism or an antiisomorphism since otherwise T^{-1} would preserve commutativity.

Let us now connect the above discussion with the zpd notion. Assume that A is a zLpd algebra and continue to assume that T is a bijective commutativity preserving linear map from A onto an algebra B. Then there is a linear map $S : A \to B$ such that (7.2) holds. Therefore, using the identity

$$[uv, w] + [wu, v] + [vw, u] = 0,$$

it follows that

$$[T(uv), T(w)] + [T(wu), T(v)] + [T(vw), T(u)] = 0$$

for all $u, v, w \in A$. Note that this can be equivalently written as

$$[B(x, y), z] + [B(z, x), y] + [B(y, z), x] = 0 \tag{7.6}$$

for all $x, y, z \in B$, where B is defined by (7.4). Thus, we have arrived at a functional identity that obviously tells us more than (7.5). Assuming, as before, that such a functional identity has only standard solutions, we obtain information on the action of T on the usual product ab of any elements $a, b \in A$, which is of course better than before when we had to deal with the Jordan product $a \circ b$.

The approach just described was taken, in the context of zLpd Banach algebras, in [49]. We will avoid stating the main result of this paper, [49, Theorem 5.3], since it involves the notion of a 3-free ring, which we did not define precisely. Instead, we will restrict our attention to the corollary concerning von Neumann algebras that are, on the one hand, zLpd (just as all C^*-algebras, see Corollary 6.12) and, on the other hand, have a sufficiently nice structure to determine when they are 3-free [3].

We need one more definition. We say that a bijective linear map $T : A \to B$ is the *direct sum of an isomorphism and an antiisomorphism* if A is the direct sum of two ideals I and J, B is the direct sum of two ideals I' and J', the restriction of T to I is an isomorphism from I onto I', and the restriction of T to J is an antiisomorphism from J onto J'. Observe that T is, in particular, a Jordan isomorphism and that both T and T^{-1} preserve commutativity.

The following is [49, Corollary 5.4].

Theorem 7.16 *Let A and B be von Neumann algebras with no central summands of type I_1 and I_2. If $T : A \to B$ is a continuous bijective commutativity preserving linear map, then either*

(a) *A contains an ideal J, not contained in $Z(A)$, such that $T(J) \subseteq Z(B)$ (and hence T^{-1} does not preserve commutativity) or*

(b) *there exist an invertible element $c \in Z(B)$, a linear map from $\mu : A \to Z(B)$, and the direct sum of an isomorphism and an antiisomorphism $\Phi : A \to B$ such that*

$$T(x) = c\Phi(x) + \mu(x) \quad (x \in A)$$

(and hence T^{-1} also preserves commutativity). Moreover, Φ and μ are continuous.

Note that Examples 7.13 and 7.15 are helpful in understanding this theorem.

Let us mention that [49] contains some other results on zLpd Banach algebras whose proofs involve functional identities.

7.3　Jordan Homomorphisms

We saw above that Jordan homomorphisms appear naturally in the study of linear maps preserving zero products or commutativity. In this section, we put them in the center of our attention and discuss the problem of describing their form. Our main result is Theorem 7.22, which has not appeared in the literature before.

Obvious examples of Jordan homomorphisms are homomorphisms and antihomomorphisms. By combining these two basic examples, like taking their direct sum as above, one obtains more complex examples of Jordan homomorphisms. Now, given a Jordan homomorphism, is it possible to express it through homomorphisms and antihomomorphisms? This question has been an active area of research since the 1950s, when it was studied in the pioneering works by Jacobson and Rickart [101], Kadison [107], and Herstein [94]. We will see that zpd algebras provide an appropriate framework for its consideration. We will work in the analytic setting, so we will actually consider zpd Banach algebras. However, it will be clear from the proofs that purely algebraic versions of our results can also be obtained.

We start with a few standard general observations. Let A and B be algebras over a field F of characteristic different from 2, and let $\Phi : A \to B$ be a Jordan homomorphism. From the identity

$$xyx = \frac{1}{2}\big(x \circ (x \circ y) - x^2 \circ y\big),$$

we deduce that

$$\Phi(xyx) = \Phi(x)\Phi(y)\Phi(x) \quad (x, y \in A). \tag{7.7}$$

Linearizing, we obtain

$$\Phi(xyz + zyx) = \Phi(x)\Phi(y)\Phi(z) + \Phi(z)\Phi(y)\Phi(x) \quad (x, y, z \in A). \tag{7.8}$$

The next identity will be obtained by computing

$$U = \Phi(xyzyx + yxzxy)$$

in two ways. Firstly, using (7.7), we obtain

$$U = \Phi\big(x(yzy)x\big) + \Phi\big(y(xzx)y\big)$$
$$= \Phi(x)\Phi(y)\Phi(z)\Phi(y)\Phi(x) + \Phi(y)\Phi(x)\Phi(z)\Phi(x)\Phi(y).$$

Secondly, (7.8) gives

$$U = \Phi\big((xy)z(yx) + (yx)z(xy)\big) = \Phi(xy)\Phi(z)\Phi(yx) + \Phi(yx)\Phi(z)\Phi(xy).$$

By comparing the two expressions, we obtain

$$\Phi(xy)\Phi(z)\Phi(yx) + \Phi(yx)\Phi(z)\Phi(xy)$$
$$= \Phi(x)\Phi(y)\Phi(z)\Phi(y)\Phi(x) + \Phi(y)\Phi(x)\Phi(z)\Phi(x)\Phi(y).$$

Using

$$\Phi(yx) = \Phi(x)\Phi(y) + \Phi(y)\Phi(x) - \Phi(xy),$$

we see that this identity can be written as

$$h(x, y)\Phi(z)a(x, y) + a(x, y)\Phi(z)h(x, y) = 0 \quad (x, y, z \in A), \tag{7.9}$$

where

$$h(x, y) = \Phi(xy) - \Phi(x)\Phi(y)$$

and

$$a(x, y) = \Phi(xy) - \Phi(y)\Phi(x).$$

Of course, $h(x, y) = 0$ for all $x, y \in A$ means that Φ is a homomorphism, and $a(x, y) = 0$ for all $x, y \in A$ means that Φ is an antihomomorphism.

Recall that an algebra B is said to be semiprime if it has no nonzero nilpotent ideals. Equivalently, for any $b \in B$, $bBb = \{0\}$ implies $b = 0$. Semiprime algebras provide a large class of algebras; in particular, *semisimple* (also called *semiprimitive*) algebras, i.e., algebras with trivial Jacobson radical, are semiprime. This includes C^*-algebras and group algebras $L^1(G)$ for a locally compact group G. The other obvious examples are prime algebras.

The next lemma is a version of [41, Corollary 2.2].

Lemma 7.17 *Let A and B be Banach algebras with B semiprime. If $\Phi : A \rightarrow B$ is a Jordan homomorphism such that $\overline{\Phi(A)} = B$, then*

$$h(x, y)Ba(z, w) = \{0\} \quad (x, y, z, w \in A) \tag{7.10}$$

and

$$h(x, y)a(z, w) = a(z, w)h(x, y) = 0 \quad (x, y, z, w \in A). \tag{7.11}$$

Proof Take $x, y \in A$, and write $h = h(x, y)$ and $a = a(x, y)$. In view of $\overline{\Phi(A)} = B$, (7.9) implies that $hua + auh = 0$ for all $u \in B$. Therefore, for all $u, z \in B$,

$$h(uaz)a = -au(azh) = (auh)za = -huaza.$$

Thus, $huaza = 0$ for all $u, z \in B$. This implies $(hua)t(hua) = 0$ for all $u, t \in B$, and hence $hua = 0$ since B is semiprime.

We have thus proved that

$$h(x, y)ua(x, y) = 0 \quad (x, y \in A, u \in B).$$

Replacing x by $x + z$, we obtain

$$h(x, y)ua(z, y) + h(z, y)ua(x, y) = 0 \quad (x, y, z \in A, u \in B).$$

Hence,

$$\big(h(x, y)ua(z, y)\big)t\big(h(x, y)ua(z, y)\big)$$
$$= -h(x, y)ua(z, y)th(z, y)ua(x, y)$$
$$\in h(x, y)Ba(x, y) = \{0\}.$$

The semiprimeness of B yields

$$h(x, y)ua(z, y) = 0 \quad (x, y, z \in A, u \in B).$$

Replacing y by $y + w$, we get

$$h(x, y)ua(z, w) + h(x, w)ua(z, y) = 0 \quad (x, y, z, w \in A, u \in B).$$

Similarly as above, we see that this gives

$$\big(h(x, y)ua(z, w)\big)t\big(h(x, y)ua(z, w)\big) = -h(x, y)ua(z, w)th(x, w)ua(z, y) = 0,$$

and so $h(x, y)ua(z, w) = 0$. This proves (7.10). Since

$$h(x, y)a(z, w)Bh(x, y)a(z, w) \subseteq h(x, y)Ba(z, w) = \{0\},$$

it follows that $h(x, y)a(z, w) = 0$. Similarly,

$$a(z, w)h(x, y)Ba(z, w)h(x, y) = \{0\}$$

implies $a(z, w)h(x, y) = 0$.   □

If B is prime, then lemma shows that Φ is either a homomorphism or an antihomomorphism. This, however, is not a new result. Its version was obtained a long time ago by Herstein [94].

In the preceding section, we have arrived at Jordan isomorphisms that are direct sums of isomorphisms and antiisomorphisms. Now we will be interested in more general, not necessarily surjective Jordan homomorphisms: $\Phi : A \to B$ is said to be the *sum of a homomorphism and an antihomomorphism* if there exist ideals I_1 and I_2 of B satisfying $I_1 I_2 = I_2 I_1 = \{0\}$, a homomorphism $\Phi_1 : A \to I_1$ and an antihomomorphism $\Phi_2 : A \to I_2$ such that

$$\Phi = \Phi_1 + \Phi_2.$$

Note that Φ is a Jordan homomorphism whose range is not necessarily a subalgebra of B.

In [41], it was shown that for every surjective Jordan homomorphism Φ from an arbitrary algebra A onto a semiprime algebra B, there exists an ideal I of A such that $I \cap J \neq \{0\}$ for every nonzero ideal J of A and the restriction of Φ to I is the sum of a homomorphism and an antihomomorphism. Example 7.19 below shows that in general $I \neq A$, that is, Φ is not always the sum of a homomorphism and an antihomomorphism on A.

A different approach, which is relevant for our purposes, was taken in [45]. We need some definitions to describe it. A Jordan homomorphism $\Phi : A \to B$ that also satisfies

$$\Phi(xyzw + wzyx) = \Phi(x)\Phi(y)\Phi(z)\Phi(w) + \Phi(w)\Phi(z)\Phi(y)\Phi(x) \qquad (7.12)$$

for all $x, y, z, w \in A$ is called a *reversal homomorphism*. Note that the sum of a homomorphism and an antihomomorphism is a reversal homomorphism. The converse, however, does not hold in general [45, Example 2.2]. Recall that the ideal of the algebra A generated by all commutators $[x, y]$, $x, y \in A$, is called the *commutator ideal* of A. By a *central ideal* of A, we mean an ideal of A contained in the center of A. We can now state [45, Theorem 4.2].

Theorem 7.18 *Let A and B be arbitrary algebras, and let $\Phi : A \to B$ be a reversal homomorphism. Suppose that the subalgebra of B generated by $\Phi(A)$ does not contain nonzero nilpotent central ideals. Then the restriction of Φ to the commutator ideal K of A is the sum of a homomorphism and an antihomomorphism.*

We continue with an example illustrating the above discussion. It is taken from [45], but it is based on the same idea as an example in [29], which the authors attributed to Kaplansky.

Example 7.19 Let $K(H)$ be the C^*-algebra of all compact operators on a separable infinite-dimensional Hilbert space H. For any $v \in K(H)$, let v^t denote the transpose of T relative to a fixed orthonormal basis. Next, let A be the algebra obtained by adjoining a unity to the algebra $K(H) \times K(H)$ by the usual process. Define $\Phi : A \to A$ by

$$\Phi((u, v) + \lambda 1) = (u, v^t) + \lambda 1 \quad (u, v \in K(H), \lambda \in \mathbb{C}).$$

Observe that Φ is a reversal homomorphism. However, Φ is not the sum of a homomorphism and an antihomomorphism. Indeed, suppose that $\Phi = \Phi_1 + \Phi_2$ as in the above definition. Set $e = \Phi_1(1)$. As Φ_1 is a homomorphism, e is an idempotent commuting with all elements in $\Phi_1(A)$. Moreover, by the definition of the sum of a homomorphism and an antihomomorphism, $e\Phi_2(A) = \Phi_2(A)e = \{0\}$, and so e commutes with all elements in A. As A has no nontrivial central idempotents, it follows that $e = 0$ or $e = 1$, yielding a contradiction that $\Phi = \Phi_2$ is an antihomomorphism or $\Phi = \Phi_1$ is a homomorphism.

The restriction of Φ to $K = K(H) \times K(H)$, however, is the sum of the homomorphism $(u, v) \mapsto (u, 0)$ and the antihomomorphism $(u, v) \mapsto (0, v^t)$. As the algebra $K(H)$ is linearly spanned by its commutators [142], K is the commutator ideal of A.

Note that this example shows that the commutator ideal may actually be the largest ideal on which a reversal homomorphism is the sum of a homomorphism and an antihomomorphism.

The next lemma, which is also taken from [45], connects Jordan homomorphisms to the topic of this book.

Lemma 7.20 *Let A and B be algebras, and let $\Phi : A \to B$ be a Jordan homomorphism. If the subalgebra of B generated by $\Phi(A)$ does not contain nonzero nilpotent central elements, then, for all $x, y \in A$,*

$$xy = yx = 0 \implies \Phi(x)\Phi(y) = 0. \tag{7.13}$$

Proof Take $x, y \in A$ such that $xy = yx = 0$. Then $x \circ y = 0$, and hence $\Phi(x) \circ \Phi(y) = 0$. Furthermore, from (7.8), we see that

$$\Phi(x)\Phi(y)\Phi(z) + \Phi(z)\Phi(y)\Phi(x) = 0$$

for every $z \in A$. Since $\Phi(y)\Phi(x) = -\Phi(x)\Phi(y)$, it follows that

$$[\Phi(x)\Phi(y), \Phi(z)] = 0.$$

This means that $c = \Phi(x)\Phi(y)$ lies in the center of the subalgebra generated by $\Phi(A)$. From (7.7), we infer that $c^2 = \Phi(xyx)\Phi(y) = 0$. By the assumption of the lemma, this gives $c = 0$.   □

The next example shows that the assumption about the non-existence of nonzero nilpotent central elements is indeed necessary.

Example 7.21 Let G_2 be the Grassmann algebra with two generators x_1 and x_2 (see Example 3.20). Thus, x_1 and x_2 satisfy

$$x_1^2 = x_2^2 = x_1 x_2 + x_2 x_1 = 0,$$

and the elements $1, x_1, x_2, x_1 x_2$ form a basis of G_2. We remark that $x_1 x_2$ lies in the center of G_2 and has square 0. Define the linear map $\Phi : G_2 \to G_2$ by

$$\Phi(1) = 1, \quad \Phi(x_1) = x_1, \quad \Phi(x_2) = x_1 x_2, \quad \Phi(x_1 x_2) = x_2.$$

One immediately checks that Φ is a Jordan automorphism. However, Φ does not satisfy (7.13). Indeed, $x_1 \cdot x_1 x_2 = x_1 x_2 \cdot x_1 = 0$, but $\Phi(x_1)\Phi(x_1 x_2) = x_1 x_2 \neq 0$.

We remark that from the formula (3.22) in Theorem 3.23, one deduces that a linear map Φ between unital algebras A and B that satisfies (7.13) and sends 1 to 1 is a Jordan homomorphism, provided that A is zpd. Note that from the preceding formula (3.21), one derives a more general identity for Φ, which does not obviously hold for all Jordan homomorphisms. The proof of the next theorem is based on this observation. However, instead of Theorem 3.23, we will use its analytic version, Theorem 6.1.

Theorem 7.22 *Let A be a zpd Banach algebra having a bounded approximate identity, let B be a semiprime Banach algebra, and let $\Phi : A \to B$ be a continuous Jordan homomorphism such that $\overline{\Phi(A)} = B$. Then Φ is a reversal homomorphism, and the restriction of Φ to the commutator ideal K of A is the sum of a homomorphism and an antihomomorphism.*

Proof Suppose an element c in the center of the subalgebra generated by $\Phi(A)$ is nilpotent. Since $\overline{\Phi(A)} = B$, c actually lies in the center of B. Hence, the ideal of B generated by c is nilpotent. As B is semiprime, it follows that $c = 0$. Lemma 7.20 therefore tells us that Φ satisfies (7.13).

We are now in a position to apply Theorem 6.1 (along with Remark 6.2) to the bilinear map $\varphi : A \times A \to B$ defined by

$$\varphi(x, y) = \Phi(x)\Phi(y).$$

Hence, we conclude that

$$\Phi(xy)\Phi(zw) + \Phi(wx)\Phi(yz) = \Phi(x)\Phi(yzw) + \Phi(wxy)\Phi(z) \tag{7.14}$$

for all $x, y, z, w \in A$.

Let $(e_\lambda)_{\lambda \in \Lambda}$ be a bounded approximate identity in A. Taking e_λ for x in (7.14), we obtain

$$\Phi(e_\lambda)\Phi(yzw) = \Phi(e_\lambda y)\Phi(zw) + \Phi(we_\lambda)\Phi(yz) - \Phi(we_\lambda y)\Phi(z) \qquad (7.15)$$

for all $y, z, w \in A$. Hence we see that the net $(\Phi(e_\lambda)\Phi(yzw))_{\lambda \in \Lambda}$ is convergent for all $y, z, w \in A$. The set of all $u \in B$ such that the net $(\Phi(e_\lambda)u)_{\lambda \in \Lambda}$ is convergent thus contains $\Phi(A^3)$ and is closed since the net $(\Phi(e_\lambda))_{\lambda \in \Lambda}$ is bounded. By Cohen's factorization theorem, the existence of a bounded approximate identity in A implies that $A^3 = A$. Hence, $(\Phi(e_\lambda)u)_{\lambda \in \Lambda}$ is convergent for every $u \in \overline{\Phi(A)} = B$. We can thus define the linear map $W : B \to B$ by

$$W(u) = \lim_{\lambda \in \Lambda} \Phi(e_\lambda)u.$$

Since the net $(\Phi(e_\lambda))_{\lambda \in \Lambda}$ is bounded, W is continuous. Our goal is to show that $(\Phi(e_\lambda))_{\lambda \in \Lambda}$ is a left approximate identity of B, that is,

$$W(u) = u \quad (u \in B) \qquad (7.16)$$

(it is also a right approximate identity, but we shall not need this).

We start the proof of (7.16) by applying (7.8) to obtain

$$\Phi(xe_\lambda y + ye_\lambda x) = \Phi(x)\Phi(e_\lambda)\Phi(y) + \Phi(y)\Phi(e_\lambda)\Phi(x)$$

for all $x, y \in A$. By taking limits, it follows that

$$\Phi(xy + yx) = \Phi(x)W(\Phi(y)) + \Phi(y)W(\Phi(x)) \quad (x, y \in A),$$

and hence

$$\Phi(x)\Phi(y) + \Phi(y)\Phi(x) = \Phi(x)W(\Phi(y)) + \Phi(y)W(\Phi(x)) \quad (x, y \in A).$$

Since $\Phi(A)$ is dense in B and W is continuous, it follows that

$$uv + vu = uW(v) + vW(u) \quad (u, v \in B);$$

that is,

$$uh(v) + vh(u) = 0 \quad (u, v \in B), \qquad (7.17)$$

where

$$h(u) = u - W(u).$$

We have to show that $h(u) = 0$ for every $u \in A$. We remark that (7.17) is a simple example of a functional identity for which the general theory [56] is applicable, but we will proceed by elementary means. Replacing v by wv, we obtain

$$uh(wv) = -(wv)h(u) = -w(vh(u)) = wuh(v) \quad (u, v, w \in B). \tag{7.18}$$

Consequently, for all $x, u, w, v \in B$, we have

$$(xu)h(wv) = w(xu)h(v),$$

and, on the other hand,

$$x(uh(wv)) = xwuh(v).$$

Comparing, we get

$$[x, w]uh(v) = 0.$$

Denoting by I the ideal of B generated by $h(B)$, we thus have

$$[B, B]BI = \{0\}.$$

In particular,

$$[I, B]B[I, B] = \{0\}.$$

Since B is semiprime, this implies that I is contained in $Z(B)$, the center of B. Hence $uh(v) \in Z(B)$, and so (7.18) can be written as

$$u(h(wv) - h(v)w) = 0 \quad (u, v, w \in B).$$

As B is semiprime and $h(v) \in Z(B)$, it follows that

$$h(wv) = h(v)w = wh(v) \quad (v, w \in B).$$

We can now rewrite (7.17) as

$$h(uv + vu) = 0 \quad (u, v \in B).$$

Replacing v by $vw + wv$ in (7.17) thus yields

$$(vw + wv)h(u) = 0 \quad (u, v, w \in B). \tag{7.19}$$

In particular, by setting $v = w = h(u)$, we get $h(u)^3 = 0$. However, the center of a semiprime algebra does not contain nonzero nilpotents, so we have $h(u) = 0$ for every $u \in B$. This proves (7.16), and so $(\Phi(e_\lambda))_{\lambda \in \Lambda}$ is a left approximate identity.

Returning to (7.15), it now follows, by taking limits, that

$$\Phi(yzw) = \Phi(y)\Phi(zw) + \Phi(w)\Phi(yz) - \Phi(wy)\Phi(z) \quad (y, z, w \in A).$$

This implies

$$\Phi(xyzw) = \Phi((xy)zw) = \Phi(xy)\Phi(zw) + \Phi(w)\Phi(xyz) - \Phi(wxy)\Phi(z)$$

and

$$\Phi(wzyx) = \Phi(wz(yx)) = \Phi(w)\Phi(zyx) + \Phi(yx)\Phi(wz) - \Phi(yxw)\Phi(z).$$

Therefore,

$$\Phi(xyzw + wzyx)$$
$$= \Phi(xy)\Phi(zw) + \Phi(yx)\Phi(wz) + \Phi(w)\Phi(xyz + zyx) - \Phi(wxy + yxw)\Phi(z).$$

Using (7.8), it follows that

$$\Phi(xyzw + wzyx)$$
$$= \Phi(xy)\Phi(zw) + \Phi(yx)\Phi(wz) \tag{7.20}$$
$$+ \Phi(w)\Phi(z)\Phi(y)\Phi(x) - \Phi(y)\Phi(x)\Phi(w)\Phi(z).$$

We now refer to Lemma 7.17, specifically to (7.11). Note that $h(x, y)a(z, w) = 0$ can be written as

$$\Phi(xy)\Phi(zw) = \Phi(x)\Phi(y)\Phi(zw) + \Phi(xy)\Phi(w)\Phi(z) - \Phi(x)\Phi(y)\Phi(w)\Phi(z),$$

and $a(y, x)h(w, z) = 0$ can be written as

$$\Phi(yx)\Phi(wz) = \Phi(x)\Phi(y)\Phi(wz) + \Phi(yx)\Phi(w)\Phi(z) - \Phi(x)\Phi(y)\Phi(w)\Phi(z).$$

Consequently,

$$\Phi(xy)\Phi(zw) + \Phi(yx)\Phi(wz)$$
$$=\Phi(x)\Phi(y)\Phi(zw + wz) + \Phi(xy + yx)\Phi(w)\Phi(z)$$
$$\quad - 2\Phi(x)\Phi(y)\Phi(w)\Phi(z)$$
$$=\Phi(x)\Phi(y)\Phi(z)\Phi(w) + \Phi(x)\Phi(y)\Phi(w)\Phi(z)$$
$$\quad + \Phi(x)\Phi(y)\Phi(w)\Phi(z) + \Phi(y)\Phi(x)\Phi(w)\Phi(z)$$
$$\quad - 2\Phi(x)\Phi(y)\Phi(w)\Phi(z)$$
$$=\Phi(x)\Phi(y)\Phi(z)\Phi(w) + \Phi(y)\Phi(x)\Phi(w)\Phi(z).$$

Together with (7.20), this gives

$$\Phi(xyzw + wzyx)$$
$$=\Phi(x)\Phi(y)\Phi(z)\Phi(w) + \Phi(y)\Phi(x)\Phi(w)\Phi(z)$$
$$\quad + \Phi(w)\Phi(z)\Phi(y)\Phi(x) - \Phi(y)\Phi(x)\Phi(w)\Phi(z)$$
$$=\Phi(x)\Phi(y)\Phi(z)\Phi(w) + \Phi(w)\Phi(z)\Phi(y)\Phi(x).$$

That is, Φ is a reversal homomorphism. By what we proved at the very beginning, the subalgebra generated by $\Phi(A)$ does not contain nonzero nilpotent central ideals. The conditions of Theorem 7.18 are thus fulfilled, and so the restriction of Φ to the commutator ideal K is the sum of a homomorphism and an antihomomorphism.   □

A somewhat different version of Theorem 7.22 for the case when A and B are C^*-algebras was obtained in [45, Theorem 5.2]. The method of proof, however, was substantially different.

It is easy to see that the algebra A is equal to its commutator ideal K if and only if A has no proper ideals I such that the quotient algebra A/I is commutative. Thus, we have the following corollary.

Corollary 7.23 *Let A be a zpd Banach algebra having a bounded approximate identity and having no nonzero commutative homomorphic images. Let B be a semiprime Banach algebra. If $\Phi : A \to B$ is a continuous Jordan homomorphism such that $\overline{\Phi(A)} = B$, then Φ is the sum of a homomorphism and an antihomomorphism.*

Corollary 7.23 is vacuous if A is commutative. But actually it is easy to consider this case. Indeed, by taking $w = x$ and $z = y$ in the identity (7.10) in Lemma 7.17, we see that for any pair of commuting elements x and y, we have

$$\big(\Phi(xy) - \Phi(x)\Phi(y)\big)B\big((\Phi(xy) - \Phi(x)\Phi(y)\big) = \{0\},$$

and hence $\Phi(xy) = \Phi(x)\Phi(y)$ since B is semiprime. This shows that if the algebra A in Theorem 7.22 is commutative, then Φ is simply a homomorphism. Thus, we can handle Jordan homomorphisms in both extreme cases, when the commutator ideal K is equal to A and when it is 0.

Derivations and Related Maps

8

There are many parallels between homomorphisms and derivations. Having just about any result on homomorphisms, one may ask whether a similar result, possibly under milder assumptions, is true for derivations. It is therefore not surprising that some of the results on homomorphisms from the preceding chapter have their counterparts for derivations. This will be shown in Sect. 8.1.

In Sects. 8.2 and 8.3, we will establish results for derivations that do not have analogues for homomorphisms. They concern local derivations and derivations whose properties are determined by quasinilpotent elements. These two topics do not seem to have any connection with zero products, so it is somewhat more surprising that they can be handled in zpd (Banach) algebras.

As in the preceding chapter, we will often only point out the main ideas and omit technical proofs.

8.1 Characterizing Derivations by Action on Zero Products

Problems on zero product preserving linear maps, studied in Sect. 7.1, have their analogues for derivation-like maps. The obvious analogy of the condition

$$xy = 0 \implies T(x)T(y) = 0$$

is the condition

$$xy = 0 \implies D(x)y + xD(y) = 0.$$

© The Author(s), under exclusive license to Springer Nature Switzerland AG 2021
M. Brešar, *Zero Product Determined Algebras*, Frontiers in Mathematics,
https://doi.org/10.1007/978-3-030-80242-4_8

Just as homomorphisms are the basic examples of linear maps satisfying the former condition, derivations are the basic examples of linear maps satisfying the latter condition. The natural question that arises is whether they are, under appropriate conditions, also the almost only examples (they are hardly literally the only examples as we will be shown in the next paragraph). The list of papers studying this and some related questions is long; in particular, it includes [2, 66, 68, 72, 102, 103, 120, 126]. Our purpose is to indicate how these questions can be handled in zpd algebras.

We will consider maps from an algebra A to an A-bimodule M. The bimodule multiplication will be denoted by $\cdot$ (so we will write $D(x) \cdot y + x \cdot D(y)$ rather than $D(x)y + xD(y)$). The set

$$Z(M) = \{c \in M \mid c \cdot x = x \cdot c \text{ for all } x \in A\}$$

is called the *center of the bimodule M*. Observe that for every $c \in Z(M)$, the linear map $D : A \to M$, $D(x) = c \cdot x$, satisfies the condition that $D(x) \cdot y + x \cdot D(y) = 0$ whenever $xy = 0$ but is not a derivation unless $D = 0$.

We begin with an analogue of Theorem 7.2.

Theorem 8.1 *Let A be an algebra, let M be an A-bimodule, and let $D : A \to M$ be a linear map with the property that for all x, $y \in A$, $xy = 0$ implies $D(x) \cdot y + x \cdot D(y) = 0$. If A is zpd, then there exists a linear map $G : A \to M$ such that*

$$G(xy) = D(x) \cdot y + x \cdot D(y) \quad (x, y \in A).$$

Moreover, if A and M are unital, then $c = D(1) \in Z(M)$ and there exists a derivation $\delta : A \to M$ such that

$$D(x) = \delta(x) + c \cdot x \quad (x \in A).$$

Proof Applying the definition of a zpd algebra (more precisely, one has to use Proposition 1.3 (iv)) to the map

$$(x, y) \mapsto D(x) \cdot y + x \cdot D(y),$$

we obtain the existence of the map G.

Assume that A and M are unital. From $G(xy) = D(x) \cdot y + x \cdot D(y)$, it then follows that

$$G(x) = D(x) + x \cdot c$$

for all $x \in A$. Similarly, $G(y) = c \cdot y + D(y)$ for all $y \in A$. Comparing both expressions, we see that $c \in Z(M)$.

Define $\delta : A \to M$ by

$$\delta(x) = D(x) - c \cdot x.$$

We have

$$
\begin{aligned}
\delta(xy) &= D(xy) - c \cdot xy \\
&= G(xy) - 2c \cdot xy \\
&= D(x) \cdot y + x \cdot D(y) - (c \cdot x) \cdot y - x \cdot (c \cdot y) \\
&= \delta(x) \cdot y + x \cdot \delta(y),
\end{aligned}
$$

which proves that δ is a derivation.   □

It is clear that the arguments of the above proof also work in the analytic setting. We therefore state the following theorem without proof.

Theorem 8.2 *Let A be a Banach algebra, let M be a Banach A-bimodule, and let $D : A \to M$ be a continuous linear map with the property that for all $x, y \in A$, $xy = 0$ implies $D(x) \cdot y + x \cdot D(y) = 0$. If A is zpd, then there exists a continuous linear map $G : A \to M$ such that*

$$G(xy) = D(x) \cdot y + x \cdot D(y) \quad (x, y \in A).$$

Moreover, if A and M are unital, then $c = D(1) \in Z(M)$ and there exists a continuous derivation $\delta : A \to M$ such that

$$D(x) = \delta(x) + c \cdot x \quad (x \in A).$$

Theorems 8.1 and 8.2 are pretty straightforward. Our purpose was just to indicate the method of proof. Some generalizations of Theorem 8.1 can be found in [35]. Assuming that A in Theorem 8.2 is not unital but only contains a bounded approximate identity, the problem becomes technically more demanding and one has to involve the second dual of M; see [6, Theorem 4.6].

Our next theorem is analogous to Theorem 7.8.

Theorem 8.3 *Let A be a prime unital algebra containing a nontrivial idempotent e. If a linear map $D : A \to A$ has the property that for all $x, y \in A$, $xy = 0$ implies $D(x)y + xD(y) = 0$, then $c = D(1) \in Z(A)$ and there exists a derivation $\delta : A \to A$ such that*

$$D(x) = \delta(x) + cx \quad (x \in A).$$

Proof As in the proof of Theorem 7.8, we let R denote the subalgebra of A generated by all idempotents in A and by I the ideal of A generated by all commutators $[e, x]$, $x \in A$. We remark that e is noncentral since the center of a prime algebra obviously cannot contain zero-divisors. Therefore, $I \neq \{0\}$.

Define $\Psi : A^4 \to A$ by

$$\Psi(x_1, x_2, x_3, x_4) = D(x_1)x_2x_3x_4 + x_1 D(x_2)x_3x_4 + x_1x_2 D(x_3)x_4 + x_1x_2x_3 D(x_4).$$

Note that, in light of Theorem 2.15 and Lemma 2.26, Lemma 7.7 tells us that

$$\Psi(u, x, y, z) = \Psi(u, x, 1, yz)$$

for all $u \in I, x, y, z \in A$. That is,

$$D(u)xyz + uD(x)yz + uxD(y)z + uxyD(z)$$
$$= D(u)xyz + uD(x)yz + uxcyz + uxD(yz).$$

We can rewrite this as

$$ux\big(D(y)z + yD(z) - cyz - D(yz)\big) = 0.$$

As A is prime, it follows that

$$D(y)z + yD(z) = cyz + D(yz) \quad (y, z \in A).$$

Setting $z = 1$ we see that $c \in Z(A)$, and hence one shows by a direct computation that $\delta : A \to A$ defined by

$$\delta(x) = D(x) - cx$$

is a derivation. $\square$

The existence of unity can be avoided, but then one has to involve the extended centroid of A; see [44, Corollary 4.6].

We continue by examining the condition

$$xy = yx = 0 \implies D(x) \bullet y + x \bullet D(y) = 0,$$

where

$$x \bullet m = x \cdot m + m \cdot x \quad (x \in A, m \in M).$$

Note the analogy with the condition

$$xy = yx = 0 \implies T(x) \circ T(y) = 0,$$

which was briefly considered in Sect. 7.1. Just like one wants to prove that the map satisfying the latter condition is close to a Jordan homomorphism, one may want to prove that the map satisfying the former condition is close to a Jordan derivation (see Definition 6.14). This seems to be somewhat easier, and Sect. 7.1 does not contain an analogue of the next theorem.

Theorem 8.4 *Let A be a unital algebra over a field of characteristic not 2, let M be a unital A-bimodule, and let $D : A \to M$ be a linear map with the property that for all $x, y \in A$, $xy = yx = 0$ implies $D(x) \bullet y + x \bullet D(y) = 0$. If A is generated by idempotents, then $c = D(1) \in Z(M)$ and there exists a Jordan derivation $\delta : A \to M$ such that*

$$D(x) = \delta(x) + c \cdot x \quad (x \in A).$$

Proof The assumption that A is generated by idempotents implies that it is zpd (Theorem 2.15). We may therefore apply Theorem 3.23 (together with Remark 3.24) to the symmetric bilinear map

$$\Psi(x, y) = D(x) \bullet y + x \bullet D(y)$$

and hence conclude that

$$\Psi(x, y) = \frac{1}{2}\Psi(x \circ y, 1),$$

that is,

$$D(x) \bullet y + x \bullet D(y) = D(x \circ y) + \frac{1}{2}(x \circ y) \bullet c. \tag{8.1}$$

Taking $x = y = e$ in (8.1) to be an idempotent, we obtain

$$2D(e) \cdot e + 2e \cdot D(e) = 2D(e) + e \cdot c + c \cdot e.$$

Multiplying from the left by e gives

$$2e \cdot D(e) \cdot e = e \cdot c + e \cdot c \cdot e,$$

while multiplying from the right gives

$$2e \cdot D(e) \cdot e = e \cdot c \cdot e + c \cdot e.$$

Comparing these two identities, we arrive at $e \cdot c = c \cdot e$. As A is generated by idempotents, this implies that $c \in Z(M)$.

From (8.1), it now follows that

$$D(x) \bullet x = D(x^2) + c \cdot x^2.$$

Hence, $\delta : A \to M$ defined as in the proof of Theorem 8.1, i.e.,

$$\delta(x) = D(x) - c \cdot x,$$

satisfies

$$\begin{aligned} \delta(x^2) =& D(x^2) - c \cdot x^2 \\ =& D(x) \bullet x - 2c \cdot x^2 \\ =& \delta(x) \bullet x, \end{aligned}$$

so δ is a Jordan derivation.   □

Corollary 8.5 *Let B be a unital algebra over a field of characteristic not 2, let $A = M_n(B)$ with $n \geq 2$, and let M be a unital A-bimodule. If a linear map $D : A \to M$ has the property that for all $x, y \in A$, $xy = yx = 0$ implies $D(x) \bullet y + x \bullet D(y) = 0$, then $c = D(1) \in Z(M)$ and there exists a derivation $\delta : A \to M$ such that*

$$D(x) = \delta(x) + c \cdot x \quad (x \in A).$$

Proof Since A is generated by idempotents (Corollary 2.4), the conclusion of Theorem 8.4 holds. What we can add here is that a Jordan derivation $\delta : A \to M$ is automatically a derivation. This was essentially proved already in 1950 by Jacobson and Rickart [101] and is explicitly stated in [43, Section 6].   □

It is clear that the analytic analogue of Theorem 8.4, involving continuous linear maps from a unital Banach algebra A that is generated by idempotents to a unital Banach A-bimodule, can be proved in essentially the same way. One just has to use Theorem 5.14 instead of Theorems 2.15 and 6.1 instead of Theorem 3.23. We omit an explicit statement and rather record a result from [7] that concerns (not necessarily unital) C^*-algebras. The idea of its proof is similar, but it is technically more involved.

A Banach A-bimodule M is said to be *essential* if it is equal to the closed linear span of the set of elements of the form $x \cdot m \cdot y$ with $x, y \in A$, $m \in M$. The following is [7, Theorem 3.1].

Theorem 8.6 *Let A be a C^*-algebra, let M be an essential Banach A-bimodule, and let $D : A \to M$ be a continuous linear map with the property that for all $x, y \in A$, $xy = yx = 0$ implies $D(x) \bullet y + x \bullet D(y) = 0$. Then there exist a continuous derivation $\delta : A \to M$ and a continuous bimodule homomorphism $H : A \to M$ such that*

$$D = \delta + H.$$

Observe that if A and M are unital, then any bimodule homomorphism $H : A \to M$ is of the form $H(x) = c \cdot x$ with $c \in Z(M)$. We also point out that the conclusion of Theorem 8.6 is that δ is a derivation. This is because every continuous Jordan derivation from A to M is a derivation [104].

8.2   Local Derivations

The main idea of this section was briefly described in Example 1.10, where we already gave the definition of a local derivation. We now state it in a more general setting that involves bimodules.

Definition 8.7 Let A be an algebra, and let M be an A-bimodule. A linear map $D : A \to M$ is called a *local derivation* if, for every $x \in A$, there exists a derivation $D_x : A \to M$ such that $D(x) = D_x(x)$.

Derivations are trivially also local derivations. The following examples show that the converse is not always true.

Example 8.8 If for every $x \in A$ there exists an $m_x \in M$ such that

$$D(x) = m_x \cdot x - x \cdot m_x,$$

then D is a local derivation (with D_x being an inner derivation). If m_x can be chosen independently of x, then D is an (inner) derivation. However, this is not always true. For example, there exist algebras (even division algebras) A such that every nonzero inner derivation from A to A is surjective [74], and hence every linear map vanishing on the center is a local derivation from A to A. Of course, not every such map is a derivation.

Example 8.9 Let A be the 3-dimensional complex unital algebra generated by a single element t satisfying $t^3 = 0$ (i.e., $A = \mathbb{C}[t]/(t^3)$). Let $D : A \to A$ be the linear map defined by $D(1) = D(t^2) = 0$ and $D(t) = t$. As $D(t)t + tD(t) = 2t^2$ and $D(t^2) = 0$, D is not a derivation. However, D is a local derivation. Indeed, take an arbitrary $x = \alpha_0 + \alpha_1 t + \alpha_2 t^2 \in A$, where $\alpha_i \in \mathbb{C}$. If $\alpha_1 = 0$, define $D_x = 0$, and if $\alpha_1 \neq 0$, define D_x by

$$D_x(\lambda_0 + \lambda_1 t + \lambda_2 t^2) = \lambda_1 t + 2(\lambda_2 - \lambda_1 \alpha_2 \alpha_1^{-1})t^2$$

for all $\lambda_i \in \mathbb{C}$. Observe that D_x is a derivation and $D(x) = D_x(x)$.

Local derivations were introduced in 1990 by Kadison [108] who proved that every continuous local derivation from a von Neumann algebra A to a dual A-bimodule is a derivation. He also provided examples of local derivations on some algebras that are not derivations. Kadison's seminal work was very influential. Over the years, many mathematicians have studied conditions implying that a local derivation is necessarily a derivation. See [42, 73, 75, 92, 105, 118, 128, 138, 148, 153, 154] where further references can be found.

It is not obvious that zero products can be useful in the study of local derivations. The following lemma reveals the connection.

Lemma 8.10 *If $D : A \to M$ is a local derivation, then for all $x, y, z \in A$,*

$$xy = yz = 0 \implies x \cdot D(y) \cdot z = 0.$$

Proof We have

$$x \cdot D(y) \cdot z = x \cdot D_y(y) \cdot z = D_y(xy) \cdot z - D_y(x) \cdot yz,$$

and so $x \cdot D(y) \cdot z = 0$ if $xy = yz = 0$.   □

The motivation for the next lemma is thus clear.

Lemma 8.11 *Let A be a unital algebra, let X be a vector space, and let $\theta : A^3 \to X$ be a trilinear map with the property that for all $x, y, z \in A$,*

$$xy = yz = 0 \implies \theta(x, y, z) = 0.$$

If a unital subalgebra R of A is a zpd algebra, then

$$\theta(r, y, s) = \theta(r, ys, 1) - \theta(1, rys, 1) + \theta(1, ry, s) \quad (y \in A, r, s \in R).$$

Proof Let $u, v \in A$ be such that $uv = 0$. Define $\theta_1 : A \times A \to X$ by

$$\theta_1(x, y) = \theta(x, yu, v).$$

If $xy = 0$, then $x \cdot yu = yu \cdot v = 0$, and hence

$$\theta_1(x, y) = \theta(x, yu, v) = 0.$$

Since R is zpd, it follows from Lemma 7.6 that θ_1 satisfies

$$\theta_1(r, y) = \theta_1(1, ry) \quad (y \in A, r \in R),$$

that is,

$$\theta(r, yu, v) - \theta(1, ryu, v) = 0 \quad (y \in A, r \in R).$$

Now fix y, r and define $\theta_2 : A \times A \to X$ by

$$\theta_2(u, v) = \theta(r, yu, v) - \theta(1, ryu, v).$$

As just observed, $uv = 0$ implies $\theta_2(u, v) = 0$. Using Lemma 7.6 again it follows that

$$\theta_2(s, 1) = \theta_2(1, s) \quad (s \in R),$$

that is,

$$\theta(r, ys, 1) - \theta(1, rys, 1) = \theta(r, y, s) - \theta(1, ry, s) \quad (y \in A, r, s \in R).$$

This is the desired conclusion. $\qquad\square$

Our first application of Lemma 8.11 concerns the basic situation where $R = A$.

Theorem 8.12 *Let A be a unital zpd algebra, and let M be a unital A-bimodule.*

(a) *If a linear map $D : A \to M$ has the property that for all $x, y, z \in A$, $xy = yz = 0$ implies $x \cdot D(y) \cdot z = 0$, then there exists a derivation $\delta : A \to M$ such that*

$$D(x) = \delta(x) + a \cdot x \quad (x \in A),$$

where $a = D(1)$.

(b) *Every local derivation from A to M is a derivation.*

Proof (a) Lemma 8.11 shows that

$$x \cdot D(y) \cdot z = x \cdot D(yz) - D(xyz) + D(xy) \cdot z \quad (x, y, z \in A).$$

Setting $y = 1$, we obtain

$$x \cdot a \cdot z = x \cdot D(z) - D(xz) + D(x) \cdot z \quad (x, z \in A).$$

Hence, the map $\delta : A \to M$ defined by

$$\delta(x) = D(x) - a \cdot x$$

satisfies

$$\begin{aligned}
\delta(xz) &= D(xz) - a \cdot xz \\
&= x \cdot D(z) + D(x) \cdot z - x \cdot (a \cdot z) - (a \cdot x) \cdot z \\
&= \delta(x) \cdot z + x \cdot \delta(z).
\end{aligned}$$

That is, δ is a derivation.

(b) If $D : A \to M$ is a local derivation, then

$$\begin{aligned}
D(1) = D_1(1) &= D_1(1^2) \\
&= D_1(1) \cdot 1 + 1 \cdot D_1(1) = 2D(1),
\end{aligned} \tag{8.2}$$

and so $D(1) = 0$. From Lemma 8.10 and (b), it follows that D is a derivation. $\qquad\square$

Remark 8.13 If A is an algebra that is generated by idempotents, then (b) holds without assuming that A and M are unital. Indeed, let $A^{\#}$ be the algebra obtained by adjoining a unity to A, make M a unital $A^{\#}$-bimodule by defining $1 \cdot m = m \cdot 1 = m$ for every $m \in M$, and extend D to $A^{\#}$ by $D(1) = 0$. Note that D is then a local derivation from $A^{\#}$ to M and that $A^{\#}$ is generated by idempotents and is hence a zpd algebra (Theorem 2.15). Therefore, (b) tells us that D is a derivation.

In the next application of Lemma 8.11, we consider the situation where R can be a proper subalgebra of A.

Theorem 8.14 *Let A be a prime algebra containing a nontrivial idempotent e. Then every local derivation from A to A is a derivation.*

Proof If A is not unital, then, as in Remark 8.13, we let $A^{\#}$ denote the algebra obtained by adjoining a unity to A, and, given a local derivation D of A, extend D to a local derivation of $A^{\#}$ by $D(1) = 0$ (we remark that $A^{\#}$ is also prime [47, Lemma 2.36], but we will not need this). If A is unital, then we set $A^{\#} = A$; as we see from (8.2), $D(1) = 0$ holds in this case too.

Let R be the subalgebra of $A^{\#}$ generated by all idempotents in $A^{\#}$. Then R is zpd (Theorem 2.15), so Lemmas 8.10 and 8.11 show that

$$rD(y)s = rD(ys) - D(rys) + D(ry)s \quad (y \in A^{\#}, r, s \in R). \tag{8.3}$$

By taking $y = 1$, we see that the restriction of D to R is a derivation.

Let I be the ideal of $A^{\#}$ generated by all commutators $[e, x]$, $x \in A$. Note that $I \neq \{0\}$ as e is nontrivial and that $I \subseteq R \cap A$ by Lemma 2.26. Take $u, v \in I$ and $y \in A$. Then $uy, v \in I \subseteq R$, and hence, since the restriction of D to R is a derivation, we have

$$D(uyv) = D(uy)v + uyD(v).$$

On the other hand, (8.3) shows that

$$D(uyv) = uD(yv) + D(uy)v - uD(y)v.$$

Comparing these two identities, we obtain

$$u\big(D(yv) - D(y)v - yD(v)\big) = 0.$$

Since A is prime and I is a nonzero ideal of A, it follows that

$$D(yv) = D(y)v + yD(v) \quad (y \in A, v \in I).$$

Consequently, for all $x, y \in A$, $v \in I$, we have, on the one hand,

$$D(xyv) = D((xy)v) = D(xy)v + xyD(v),$$

and, on the other hand,

$$D(xyv) = D(x(yv)) = D(x)yv + xD(yv)$$
$$= D(x)yv + xD(y)v + xyD(v).$$

By comparing, we obtain

$$\big(D(xy) - D(x)y - xD(y)\big)v = 0 \quad (x, y \in A, v \in I).$$

As above, the primeness of A yields $D(xy) = D(x)y + xD(y)$ for all $x, y \in A$. That is, D is a derivation.   □

The last two results were proved, in a slightly different manner, in [44].

Let us turn to the analytic context. Note that by making some obvious modifications in the proof of Theorem 8.12, we obtain the following theorem.

Theorem 8.15 *Let A be a unital zpd Banach algebra, and let M be a unital Banach A-bimodule.*

(a) *If a continuous linear map $D : A \to M$ has the property that for all $x, y, z \in A$, $xy = yz = 0$ implies $x \cdot D(y) \cdot z = 0$, then there exists a continuous derivation $\delta : A \to M$ such that*

$$D(x) = \delta(x) + a \cdot x \quad (x \in A),$$

where $a = D(1)$.

(b) *Every continuous local derivation from A to M is a derivation.*

Without assuming that A and M are unital, the problem of course becomes more complicated. See [6, Theorem 4.5] for a non-unital version of (a). With respect to (b), it is clear from the argument in Remark 8.13 that the condition we need is that the Banach algebra $A^{\#}$, obtained by adjoining a unity to the Banach algebra A, is zpd. Proposition 1.19 indicates that, unfortunately, this condition is rather delicate. However, if A is a C^*-algebra, then so is $A^{\#}$ and is hence a zpd Banach algebra (Theorem 5.19). We thus have the following corollary.

Corollary 8.16 *Every continuous local derivation from a C^*-algebra A to a Banach A-bimodule M is a derivation.*

Corollary 8.16 was first proved by Johnson [105] (who also proved that local derivations from A to M are automatically continuous, so the continuity assumption is actually redundant). Our method of proof is entirely different. For C^*-algebras, this method was discovered in [4], which is one of the crucial papers from which the zpd concept arose.

The ideas presented above have led to more profound analytic studies; see [11, 12, 64, 149–151, 157].

Let us finally mention that local automorphisms have also been extensively studied in the literature. However, it is not clear whether the zpd condition can be useful in their treatment.

On the other hand, using the methods presented above certain problems on homomorphism-like maps can be studied. Specifically, one can prove that, under suitable assumptions, a linear map T from an algebra A to an algebra B that satisfies

$$xy = yz = 0 \implies T(x)T(y)T(z) = 0$$

is close to a homomorphism; see [4, 44].

8.3 Derivations and Quasinilpotent Elements

Let A be a Banach algebra. A *continuous representation* of A on a Banach space X is a nontrivial continuous homomorphism π from A to the Banach algebra $\mathscr{B}(X)$ of all continuous linear operators on X. We say that π is *irreducible* if $\{0\}$ and X are the only subspaces of X that are invariant under every operator in $\pi(A)$. This is equivalent to the condition that $\pi(A)\xi = X$ for every $\xi \neq 0$ in X. Indeed, this is because the space $\pi(A)\xi$ is obviously invariant under all operators in $\pi(A)$ and so is the space of all $\xi \in X$ such that $\pi(A)\xi = \{0\}$. Note that a character is a continuous irreducible representation on the 1-dimensional Banach space $\mathbb{C}$.

The (Jacobson) *radical* of a Banach algebra A, rad(A), is the intersection of the kernels of all continuous irreducible representations of A; if there are no continuous irreducible representations of A, we define rad$(A) = A$. A Banach algebra is said to be *semisimple* if rad$(A) = \{0\}$.

Let $a \in A$. Assuming that A is unital, we define the *spectrum* of a as the set

$$\sigma(a) = \{\lambda \in \mathbb{C} \mid a - \lambda 1 \text{ is not invertible}\}$$

and the *spectral radius* of a as

$$r(a) = \sup\{|\lambda| \mid \lambda \in \sigma(a)\}.$$

If a is not unital, then we define $\sigma(a)$ and $r(a)$ by considering a as an element of the Banach algebra $A^{\#}$ obtained by adjoining a unity to A. We say that a is *quasinilpotent* if $r(a) = 0$. It is easy to see that a nilpotent element is also quasinilpotent. We denote the set of all quasinilpotent elements in A by $Q(A)$. It is well-known that rad$(A) \subseteq Q(A)$. Moreover, if $a \in A$ is such that $Aa \subseteq Q(A)$ (or $aA \subseteq Q(A)$), then $a \in$ rad(A).

In 1955, Singer and Wermer proved a famous theorem that states that the image of a continuous derivation from a commutative Banach algebra A to itself always lies in rad(A) [156]. They conjectured that the assumption of continuity is unnecessary, which was confirmed more than 30 years later by Thomas [160]. Another line of investigation stemming from the Singer–Wermer theorem, and the one in which we are now interested, is finding conditions implying that a derivation of a noncommutative Banach algebra A maps A into rad(A).

In [122], Le Page proved that if δ is an inner derivation of a Banach algebra A, then $\delta(A) \subseteq Q(A)$ implies $\delta(A) \subseteq \mathrm{rad}(A)$. This was generalized to arbitrary derivations by Turovskii and Shulman [162] (somewhat later, but independently, this was also established in [133]). In [51], it was shown that if a derivation δ of A satisfies $r(\delta(x)) \leq Mr(x)$ for all $x \in A$, where $M > 0$ is a constant, then $\delta(A) \subseteq \mathrm{rad}(A)$. Furthermore, in [114] it was shown that if $\delta(x) = [q, x]$ is an inner derivation with $q \in Q(A)$, then $\delta(Q(A)) \subseteq Q(A)$ implies $\delta(A) \subseteq \mathrm{rad}(A)$.

There are many more results stating that under appropriate assumptions a derivation maps a Banach algebra A into $\mathrm{rad}(A)$ (some older ones are surveyed in [132] and [136]). We mentioned only those that consider conditions that are special cases of the condition that a derivation δ leaves the set $Q(A)$ invariant. Does this condition itself imply that $\delta(A)$ is contained in $\mathrm{rad}(A)$? In general, the answer is negative. For example, if A is a noncommutative Banach algebra in which 0 is the only quasinilpotent element [79], then every nonzero inner derivation of A gives rise to a counterexample. In zpd Banach algebras, however, the answer is positive. Thus, the following theorem, established in [14], holds.

Theorem 8.17 *Let A be a zpd Banach algebra. If a derivation δ of A satisfies $\delta(Q(A)) \subseteq Q(A)$, then $\delta(A) \subseteq \mathrm{rad}(A)$.*

We will not give a full proof but just explain its idea. The key is the following lemma.

Lemma 8.18 *Let A be a zpd Banach algebra, and let π be a continuous irreducible representation of A on a Banach space X with $\dim X \geq 2$. Then there exists a $q \in A$ such that $q^2 = 0$ and $\pi(q) \neq 0$.*

Proof It is enough to prove that there exist $x, y \in A$ such that

$$\pi(x) \neq 0, \ \pi(y) \neq 0, \ \text{and } xy = 0. \tag{8.4}$$

Indeed, if x and y satisfy (8.4) and $\xi_1, \xi_2 \in X$ are such that $\pi(x)\xi_1 \neq 0$ and $\pi(y)\xi_2 \neq 0$, then by choosing $z \in A$ satisfying $\pi(z)(\pi(x)\xi_1) = \xi_2$ (such exists by irreducibility of π), we see that $q = yzx$ satisfies $q^2 = 0$ and $\pi(q)\xi_1 \neq 0$.

Assume toward a contradiction that x and y satisfying (8.4) do not exist. Thus, for all $x, y \in A$, $xy = 0$ implies $\pi(x) = 0$ or $\pi(y) = 0$. That is, the continuous bilinear map $\varphi: A \times A \to \mathscr{B}(X) \widehat{\otimes} \mathscr{B}(X)$ defined by

$$\varphi(x, y) = \pi(x) \otimes \pi(y)$$

has the property that $\varphi(x, y) = 0$ whenever $xy = 0$ (here, as always, $\widehat{\otimes}$ stands for the projective tensor product). Since A is zpd, we have (by Proposition 4.9 (iv)) $\varphi(xy, z) = \varphi(x, yz)$ for all $x, y, z \in A$, i.e.,

$$\pi(x)\pi(y) \otimes \pi(z) = \pi(x) \otimes \pi(y)\pi(z) \quad (x, y, z \in A).$$

This implies that $\pi(z)$ and $\pi(y)\pi(z)$ are linearly dependent for any $y, z \in A$. However, this is impossible since $\dim X \geq 2$. Indeed, take linearly independent $\xi, \eta \in X$. By irreducibility of π, there are $y, z \in A$ such that $\pi(y)\xi = \eta$ and $\pi(z)\xi = \xi$. Hence, $\pi(y)\pi(z)\xi = \eta$, and so $\pi(z)$ and $\pi(y)\pi(z)$ are linearly independent. $\square$

A sketch of proof of Theorem 8.17 is as follows. The goal is to show that every continuous irreducible representation π of A on a Banach space X vanishes on $\delta(A)$. The first easy step is the reduction to the case where A is semisimple and unital. Then δ is continuous [106] and leaves the kernel of π invariant [155]. The latter makes the case where $\dim X = 1$ trivial. We may therefore assume that $\dim X \geq 2$, and so, by Lemma 8.18, A contains square-zero elements q such that $\pi(q) \neq 0$, and, by the assumption of the theorem, $\delta(q) \in Q(A)$. Showing that this implies that $\pi(\delta(A)) = \{0\}$ is the technically most involved step, based on the Jacobson density theorem and its generalizations. Specifically, these are the density theorem for invertible elements (see, e.g., [27, Corollary 4.2.6]) and the density theorem for outer derivations from [31,52]. The latter theorem states that, for any continuous derivation δ of a Banach algebra A and any continuous irreducible representation π of A on a Banach space X, the condition that there does not exist a continuous linear operator $T : X \to X$ such that

$$\pi(\delta(x)) = T\pi(x) - \pi(x)T \quad (x \in A) \tag{8.5}$$

is equivalent to the condition that for any linearly independent $\xi_1, \ldots, \xi_n \in X$ and any $\eta_1, \ldots, \eta_n, \zeta_1, \ldots, \zeta_n \in X$, there exists an $x \in A$ such that

$$\pi(x)\xi_i = \eta_i \quad \text{and} \quad \pi(\delta(x))\xi_i = \zeta_i$$

for $i = 1, \ldots, n$. One thus has to treat separately two cases: the one where π is such that (8.5) holds for some T and the one where there is not such T.

Theorem 8.17 is actually true for every Banach algebra A for which there exists a family of continuous irreducible representations $(\pi_i)_{i \in I}$ of A on Banach spaces X_i satisfying the following two conditions:

(a) $\bigcap_i \ker \pi_i = \mathrm{rad}(A)$.
(b) If $\dim X_i \geq 2$, then there exists a $q \in A$ such that $q^2 = 0$ and $\pi_i(q) \neq 0$.

Besides zpd Banach algebras, this class of Banach algebras also includes all commutative Banach algebras and Banach algebras $\mathscr{B}(X)$ for any Banach space X (which are not always zpd, see Example 5.26). Also, from the proof of Theorem 8.17, as given in [14], it is clear that if A is semisimple, then the assumption that $\delta(Q(A)) \subseteq Q(A)$ can be replaced by a milder assumption that $\delta(q) \subseteq Q(A)$ for every square-zero element $q \in A$.

Our main examples of zpd Banach algebras, C^*-algebras, and group algebras $L^1(G)$, are all semisimple. The following corollaries of Theorem 8.17 therefore hold.

Corollary 8.19 *Let A be a C^*-algebra. If a derivation δ of A satisfies $d(Q(A)) \subseteq Q(A)$, then $\delta = 0$.*

Corollary 8.20 *Let G be a locally compact group, and let $A = L^1(G)$. If a derivation δ of A satisfies $d(Q(A)) \subseteq Q(A)$, then $\delta = 0$.*

We also mention the paper [125] in which Theorem 8.17 is extended to generalized derivations.

Let us now turn to a similar yet different topic. First we provide some motivation.

We continue to assume that A is a Banach algebra. As already mentioned, an element $a \in A$ belongs to rad(A) if and only if $aA \subseteq Q(A)$. Thus, if A is semisimple, then $aA \subseteq Q(A)$ implies $a = 0$. Since the spectral radius is submultiplicative on commuting elements, every element a from the center $Z(A)$ satisfies $r(ax) \leq Mr(x)$, $x \in A$, where $M = r(a)$. If A is semisimple, then the converse is also true, i.e., $r(ax) \leq Mr(x)$, $x \in A$, implies $a \in Z(A)$ [40, 144]. A more intriguing problem is finding connections between elements $a, b \in A$ satisfying $r(ax) \leq r(bx)$, $x \in A$. This was studied in the case where A is a C^*-algebra in [13, 55].

One can study similar problems for derivations. By the result already mentioned, $\delta(A) \subseteq Q(A)$ implies $\delta(A) \subseteq$ rad(A) for every derivation δ of A [162]. In particular, if A is semisimple, then $a \in Z(A)$ if and only if $[a, A] \subseteq Q(A)$. Furthermore, [59] considers, among other things, the condition $r([a, x]) = r([b, x])$ for all x in a C^*-algebra A.

The following theorem considers more general conditions than those just discussed. It can be extracted from the arguments in [13, 55, 59]. By $\{b\}^c$ we denote the *commutant* of $\{b\}$, i.e., the set of elements in A that commute with b. Note that $\{b\}^c$ is a closed subalgebra. We will also deal with the *second commutant*, $\{b\}^{cc}$, which consists of all elements that commute with every element in $\{b\}^c$.

Theorem 8.21 *Let A be a semisimple unital Banach algebra. Suppose that $a, b \in A$ satisfy one of the following two conditions:*

(a) *For every $x \in A$, $bx \in Q(A)$ implies $ax \in Q(A)$.*
(b) *For every $x \in A$, $[b, x] \in Q(A)$ implies $[a, x] \in Q(A)$.*

If $\{b\}^c$ is a zpd Banach algebra, then $a \in \{b\}^{cc}$.

Proof Set $B = \{b\}^c$. Suppose $x, y \in B$ are such that $xy = 0$. As x and y commute with b, we then also have $xby = 0$. Let us show that each of the two conditions, (a) and (b), implies that $xay = 0$ too.

Suppose first that (a) holds. Since

$$(byux)^2 = byu(xby)ux = 0 \quad (u \in A),$$

$byux \in Q(A)$. Hence, by our assumption, $ayux \in Q(A)$. Since the spectral radius satisfies $r(zw) = r(wz)$ for all $z, w \in A$, it follows that $xayu \in Q(A)$. As u here is an arbitrary element of A and A is semisimple, it follows that $xay = 0$, as desired.

Assume now that (b) holds. From $xy = xby = 0$, we obtain

$$[b, yux]^2 = -yuxb^2yux \quad (u \in A),$$

which further implies that $[b, yux]^3 = 0$ for all $u \in A$. Therefore, $[b, yux] \in Q(A)$, and hence, by our assumption,

$$[a, yux] \in Q(A) \quad (u \in A). \tag{8.6}$$

Suppose $xay \neq 0$. Since A is semisimple, there exists a continuous irreducible representation π of A on a Banach space X such that $\pi(xay) \neq 0$. Take $\xi \in X$ such that $\pi(xay)\xi \neq 0$. Since π is irreducible, there exists a $u \in A$ such that $\pi(u)\pi(xay)\xi = -\xi$. Then $\eta = \pi(y)\xi \neq 0$ and

$$\pi([a, yux])\eta = \pi(a)\pi(yu)\pi(xy)\xi - \pi(y)\pi(u)\pi(xay)\xi = \eta.$$

Thus, 1 lies in the spectrum of $\pi([a, yux])$ and hence $r([a, yux]) \geq 1$, which contradicts (8.6).

We have thus shown that in each of the two cases, (a) and (b), the continuous bilinear map $\varphi : B \times B \to A$ defined by

$$\varphi(x, y) = xay$$

has the property that $xy = 0$ implies $\varphi(x, y) = 0$. Since B is assumed to be zpd, it follows that $\varphi(x, 1) = \varphi(1, x)$ for all $x \in B$. This means that a commutes with every element in B, i.e., $a \in \{b\}^{cc}$. $\square$

Some related conditions can also be considered. For example, the condition that for every $x \in A$, $b \circ x \in Q(A)$ implies $a \circ x \in Q(A)$ can be handled in almost the same way as (b).

In general, it may be difficult to check whether $\{b\}^c$ is a zpd Banach algebra. The following corollary provides one situation where this holds. Recall first that an element a

in a C^*-algebra A is said to be *normal* if $aa^* = a^*a$. *Fuglede's theorem* states that if a normal element $a \in A$ commutes with $b \in A$, then so does a^*.

Corollary 8.22 *Let A be a unital C^*-algebra. Suppose that $a, b \in A$ satisfy one of the following two conditions:*

(a) *For every $x \in A$, $bx \in Q(A)$ implies $ax \in Q(A)$.*
(b) *For every $x \in A$, $[b, x] \in Q(A)$ implies $[a, x] \in Q(A)$.*

If b is normal, then a is normal and lies in $\{b\}^{cc}$.

Proof Since b is normal, $\{b\}^c$ is a C^*-algebra by Fuglede's theorem and is hence zpd (Theorem 5.19). Theorem 8.21 therefore tells us that $a \in \{b\}^{cc}$. In particular, $ab = ba$, and hence, since b is normal, $a^*b = ba^*$. Thus, $a^* \in \{b\}^c$, and therefore $aa^* = a^*a$. □

The papers [13, 55, 59] contain some deeper and more definitive results ([55] also presents applications to some preserves problems). For example, [13, Theorem 3.6] states that the condition that elements a and b in a unital C^*-algebra satisfy $r(ax) \leq r(bx)$ for all $x \in A$ is equivalent to the condition that there exists a $z \in Z(A'')$ such that $a = zb$ and $\|z\| \leq 1$. A certain version of the (a) part of Corollary 8.22 is a crucial auxiliary result needed for its proof.

Many other results in this line of research have been obtained over the recent years [38, 39, 62, 152, 161]. However, we will not discuss them here since their proofs are not related to the zpd concept.

Miscellany

In this last chapter, we consider three unrelated topics in which the zpd property has turned out useful. The first topic concerns the connection between commutators and elements having square zero or having some related property. The second topic concerns conditions on n-linear functionals that arose from the study of orthogonally additive polynomials. The last and third topic concerns nonassociative products on the linear space of square matrices that are related to the standard product in the Lie-theoretic sense.

Like in the other two chapters of Part III, we will sometimes omit technical details.

9.1   Commutators and Special-Type Elements

Let R be a noncommutative ring. Is every commutator $[x, y]$ of two elements x and y in R a sum of nilpotent elements? The answer, of course, depends on R. If R has no nonzero nilpotent elements, then it is obviously negative. On the other hand, it is positive if R is the matrix ring $M_n(S)$, where S is any ring. This was established in the 1958 paper by Harris [93, Proposition 1] who, however, attributes this result to Kaplansky. As the matrix ring serves as a model of a noncommutative ring, this suggests that the above question is natural and interesting. In 1961, Herstein asked whether the answer is positive in any simple ring R with zero-divisors [96, Problem 8 (d)]. It took almost 50 years before the answer was given. In 2010, Chebotar, Lee, and Puczylowski showed that it is negative in general [71, Theorem 11], but positive if R contains a nontrivial idempotent [71, Theorem 6]—moreover, in this case every commutator in R is a sum of square-zero (not only nilpotent) elements. We remark that a simple ring containing one nontrivial idempotent is actually generated by its idempotents—see Lemma 2.26.

© The Author(s), under exclusive license to Springer Nature Switzerland AG 2021
M. Brešar, *Zero Product Determined Algebras*, Frontiers in Mathematics,
https://doi.org/10.1007/978-3-030-80242-4_9

We already pointed out in Example 1.11 that the zpd property is applicable to the problem just presented. The next simple theorem is merely a reinterpretation of this example.

Theorem 9.1 *Let A be a zpd algebra such that $A^2 = A$. Then every commutator in A is a sum of square-zero elements in A.*

Proof Let N be the linear span of all square-zero elements in A (equivalently, N is the set of all sums of square-zero elements). Define the bilinear map φ from $A \times A$ to the quotient vector space A/N by

$$\varphi(x, y) = yx + N.$$

If $xy = 0$, then $(yx)^2 = 0$, and hence $\varphi(x, y) = 0$. As A is zpd, $\varphi(xy, z) = \varphi(x, yz)$ for all $x, y, z \in A$; that is,

$$[zx, y] = z(xy) - (yz)x \in N.$$

Since $A^2 = A$, this shows that every commutator lies in N.   □

Together with Theorem 2.15, this theorem yields the following corollary.

Corollary 9.2 *If a unital algebra A is generated by idempotents, then every commutator in A is a sum of square-zero elements in A.*

A careful inspection of the proof shows that Corollary 9.2 holds for unital rings, not only for unital algebras.

Remark 9.3 A bit more can be said for a unital zpd algebra A. Recall from Proposition 2.7 that for every $w \in A$, there exist $x_i, y_i \in A, i = 1, \ldots, n$, such that $x_i y_i = 0$ for each i and

$$1 \otimes w - w \otimes 1 = \sum_{i=1}^{n} x_i \otimes y_i.$$

Applying the flip map $x \otimes y \mapsto y \otimes x$, it follows that

$$w \otimes 1 - 1 \otimes w = \sum_{i=1}^{n} y_i \otimes x_i.$$

Next, we use the linear map $x \otimes y \mapsto L_x R_y$ from $A \otimes A$ to the multiplication algebra $M(A)$ to obtain

$$L_w - R_w = \sum_{i=1}^{n} L_{y_i} \otimes R_{x_i}.$$ (9.1)

Thus,

$$[w, z] = \sum_{i=1}^{n} y_i z x_i$$

for every $z \in A$. Observe that $x_i y_i = 0$ implies $(y_i z x_i)^2 = 0$. We have thus arrived at the conclusion that every commutator in A is a sum of square-zero elements. However, we proved more than that. Indeed, (9.1) shows that every inner derivation of A is a sum of operators $L_{y_i} R_{x_i}$ whose ranges consist of square-zero elements. If A is generated by idempotents, then Remark 2.16 indicates how to derive explicit formulas for x_i and y_i.

We proceed to the analytic results. It will be quite clear that some of our arguments lead to similar algebraic results, but we will not state them explicitly. In what follows, we will survey certain parts of the papers [17] and [20].

It should be mentioned, first of all, that the question whether every commutator is a sum of square-zero elements has also been extensively studied in operator algebras. A positive answer was first obtained for the algebra $\mathcal{B}(H)$, where H is a Hilbert space [83, 140, 164] and was later extended to various more general C^*-algebras; see, e.g., [113, 130, 131, 141].

The next theorem can be proved in essentially the same way as Theorem 9.1. The only difference is that now one takes N to be the closed linear span of all square-zero elements (see the proof of Theorem 6.23).

Theorem 9.4 *Let A be a zpd Banach algebra such that $\overline{A^2} = A$. Then every commutator in A lies in the closed linear span of square-zero elements in A.*

This theorem of course applies to C^*-algebras and group algebras $L^1(G)$ for any locally compact group G, and so it generalizes and unifies two classical results, which state that, provided that they are not commutative, algebras from these two classes contain nonzero square-zero elements. For C^*-algebras, this was proved by Kaplansky (see, e.g., [109, p. 292]) and for group algebras by Behncke [33].

Theorem 9.4 obtains a nicer form for C^*-algebras, since their square-zero elements are always commutators [113].

Corollary 9.5 *If A is a C^*-algebra, then the closed linear span of all square-zero elements in A coincides with the closed linear span of all commutators in A.*

Using different methods, this corollary was also proved by Robert [146, Corollary 2.3].

An obvious consequence of Theorem 9.4 is that the closed commutator ideal of A (i.e., the closed ideal generated by all commutators in A) is contained in the closed ideal generated by all square-zero elements of A. The following corollary tells us that, under an additional assumption on A, this can be improved. For proof, see [17, Theorem 5.3].

Corollary 9.6 *Let A be a zpd Banach algebra such that $\overline{A^2} = A$. Suppose that the quotient algebra A/I is semisimple for each proper closed ideal I of A. Then the closed commutator ideal of A is equal to the closed subalgebra generated by all square-zero elements in A.*

Examples of Banach algebras satisfying the conditions of Corollary 9.6 are C^*-algebras and group algebras $L^1(G)$ for a compact group G (by [112, Theorem 15]).

Unfortunately, the method of proof of Theorem 9.4 does not bring any information about the connection between commutators and the (non-closed) linear span of square-zero elements. A different approach is needed for this. Another more subtle problem that has been considered in this line of research asks whether, for a given algebra A, there exists a positive integer N such that every commutator in A is a sum of at most N square-zero elements. It has turned out that, for example, $N = 4$ if $A = M_n(F)$, where F is a field [77] or if $A = \mathscr{B}(H)$, where H is a Hilbert space [164], $N = 22$ if $A = M_n(B)$, where B is any unital algebra [17], etc. The results of this kind have turned out to be applicable to the Waring problem for noncommutative polynomials in associative algebras [50].

One can say more about the form of square-zero elements in a noncommutative zpd Banach algebra A. Assuming that $\overline{A^2} = A$, it is easy to see that A contains a square-zero element of the form yx, where $x, y \in A$ are such that $xy = 0$. Indeed, one just has to consider the map $\varphi : A \times A \to A$ given by $\varphi(x, y) = yx$. The following result [17, Theorem 6.5] is slightly more sophisticated.

Theorem 9.7 *Let A be a semiprime zpd Banach algebra. If $a \in A$ does not lie in the center $Z(A)$, then there exist $x, y \in A$ such that $xy = yx = 0$ and $xay \neq 0$.*

Every noncentral element a thus generates a special square-zero element xay, and so, in particular, every ideal that is not contained in $Z(A)$ has an abundance of square-zero elements (see [129, Corollary 2.8] for a related result for C^*-algebras).

We give the proof of Theorem 9.7 only for the case where A is unital. Consider the map $\varphi : A \times A \to A$, $\varphi(x, y) = xay$. We have to show that there exist $x, y \in A$ such that $xy = yx = 0$ and $\varphi(x, y) \neq 0$. If this was not true, then it would follow from Theorem 6.1 that

$$\varphi(y, w) + \varphi(w, y) = \varphi(1, yw) + \varphi(wy, 1) \quad (y, w \in A),$$

and hence $[[y, a], w] = 0$. That is, $[y, a] \in Z(A)$ for all $y \in A$. Replacing y by ya, it follows that $[y, a]a \in Z(A)$, and hence $[y, a]^2 = [y, [y, a]a] = 0$ for all $y \in A$. Note that the center of a semiprime algebra does not contain nonzero nilpotent elements. Therefore, $[y, a] = 0$ for every $y \in A$, a contradiction.

We conclude this section by stating three rather simple results that concern zJpd Banach algebras and anticommuting elements. For any Banach algebra A, we set

$$A^\circ = \{xy \mid x, y \in A, x \circ y = 0\}.$$

It can easily occur that $A^\circ = \{0\}$; for example, if A is commutative, then this is certainly the case. To give a nontrivial example, take a noncentral idempotent e in a unital Banach algebra A. Then every element of the form $ea(1 - e)$, $a \in A$, lies in A°. Indeed, $x = ea(1 - e)$ and $y = 1 - 2e$ satisfy $ea(1 - e) = xy$ and $x \circ y = 0$.

The next theorem shows that A° is a large subset of any noncommutative zJpd Banach algebra A, even though such a Banach algebra does not necessarily contain nontrivial idempotents.

Theorem 9.8 *Let A be a zJpd Banach algebra. Then every commutator in A lies in the closed linear span of A°.*

Proof Let M denote the closed linear span of A°. Define $\varphi \colon A \times A \to A/M$ by

$$\varphi(x, y) = xy + M.$$

Clearly, φ is a continuous bilinear map with the property that $x \circ y = 0$ implies $\varphi(x, y) = 0$. Since A is zJpd, it follows that φ is, in particular, symmetric. This shows that $[x, y] \in M$ for all $x, y \in A$. $\square$

The following corollary is interesting in light of Theorem 6.23.

Corollary 9.9 *Let A be a weakly amenable zpd Banach algebra having a bounded approximate identity. Then A is a zJpd Banach algebra if and only if every commutator in A lies in the closed linear span of A°.*

Proof We only have to prove the "if" part. Thus, assume that all commutators lie in the closed linear span of A°, and let $\varphi \colon A \times A \to \mathbb{C}$ be a continuous bilinear functional such that $\varphi(x, y) = 0$ whenever $x \circ y = 0$. Since $xy = yx = 0$ implies $x \circ y = 0$, Theorem 6.6 shows that there exist $\tau_1, \tau_2 \in A'$ such that

$$\varphi(x, y) = \tau_1(xy) + \tau_2(yx) \quad (x, y \in A).$$

If $x, y \in A$ are such that $x \circ y = 0$, then $\varphi(x, y) = 0$, and hence $(\tau_1 - \tau_2)(xy) = 0$. Thus, the continuous linear functional $\tau_1 - \tau_2$ vanishes on A°, and hence it also vanishes on all commutators. Consequently,

$$\varphi(x, y) = \frac{1}{2}(\tau_1 + \tau_2)(x \circ y) + \frac{1}{2}(\tau_1 - \tau_2)([x, y])$$

$$= \frac{1}{2}(\tau_1 + \tau_2)(x \circ y),$$

which proves that A is zJpd.   □

We conclude with a result similar to Theorem 9.7.

Theorem 9.10 *Let A be a semiprime zJpd Banach algebra. If $a \in A$ does not lie in the center $Z(A)$, then there exist $x, y \in A$ such that $x \circ y = 0$ and $x \circ (y \circ a) \neq 0$.*

As one would expect, the proof is based on considering the map $\varphi : A \times A \to A$ defined by

$$\varphi(x, y) = x \circ (y \circ a).$$

Assuming that the theorem is not true, it follows that this map is symmetric, which results in $[a, [x, y]] = $ for all $x, y \in A$. In a semiprime algebra, this is enough to conclude that $a \in Z(A)$, which contradicts the assumption (see [20, Proposition 4.1] for details).

9.2   Orthogonality Conditions on n-Linear Maps

Let X and Y be vector spaces. A map $P : X \to Y$ is said to be an *n-homogeneous polynomial* if there exists an n-linear map $\varphi : X^n \to Y$ such that

$$P(x) = \varphi(x, \ldots, x) \quad (x \in X).$$

If the characteristic of the underlying field is 0, we may assume that φ is symmetric, i.e.,

$$\varphi(x_1, \ldots, x_n) = \varphi(x_{\sigma(1)}, \ldots, x_{\sigma(n)}) \quad (x_1, \ldots, x_n \in X)$$

for every permutation $\sigma \in S_n$, since otherwise we replace it by

$$\frac{1}{n!} \sum_{\sigma \in S_n} \varphi(x_{\sigma(1)}, \ldots, x_{\sigma(n)}).$$

Assume now that $X = A$ is an algebra. We say that an n-homogeneous polynomial $P : A \to Y$ is *orthogonally additive* if for all $x, y \in A$,

$$xy = yx = 0 \implies P(x + y) = P(x) + P(y).$$

If A is commutative, then this condition can be written more simply as

$$xy = 0 \implies P(x + y) = P(x) + P(y).$$

An obvious example of an orthogonally additive n-homogeneous polynomial is any map of the form

$$P(x) = T(x^n) \quad (x \in A), \tag{9.2}$$

where $T : A \to Y$ is a linear map. Is every orthogonally additive n-homogeneous polynomial of such a form? If $n = 2$ and A is zpd, then Theorem 3.23 (and Theorem 6.1 in the analytic context) shows that, under mild additional assumptions, the answer is positive. For $n > 2$, this is no longer clear.

The question whether orthogonally additive n-homogeneous polynomials can be presented in the form (9.2) has roots in the Banach lattice theory [36, 159]. Over the last 15 years, it has been extensively studied in Banach algebras where one usually assumes that the involved maps are continuous. A positive answer was obtained for various classes of Banach algebras, including C^*-algebras (first commutative [36, 65, 143] and then general [139]), commutative Banach algebras with property $\mathbb{A}$ (i.e., commutative Banach algebras satisfying the condition of Theorem 5.20) and having a bounded approximate identity [16], group algebras $L^1(G)$, where G is a compact group [21], and Banach algebras of approximable operators [22]. We could list other Banach algebras, but we restricted ourselves to those that are zpd.

The obvious problem that emerges is whether the answer to the above question is positive in any zpd Banach algebra. We do not know this at present. It seems, however, that it is not easy to reduce the general case to the case where $n = 2$. In [163, Section 2], Villena actually did make this reduction, but only for some special commutative Banach algebras. Using it, he applied the fact that C^*-algebras are zpd to give another proof that the answer to this question is positive for commutative C^*-algebras. Also, he showed that the answer is positive for the Banach algebra of Lipschitz functions $\mathrm{lip}_\alpha(K)$, where $0 < \alpha < 1$ and K is a compact metric space, using the fact that this Banach algebra is symmetrically zpd (see Remark 6.21).

Thus, there is at least one direct application of the zpd theory to the problem of orthogonal additivity. We will not discuss it in greater detail, but rather examine some related "orthogonality conditions," which are interesting by themselves and also shed some light on orthogonally additive polynomials. In order to introduce them, we first reformulate the definition of these polynomials.

We will work in the analytic setting, although our arguments work in the algebraic setting as well. Also, for simplicity of exposition, we consider only n-linear functionals (i.e., the case where $Y = \mathbb{C}$). This restriction is not essential and is used in many of the aforementioned papers.

Thus, let A be a Banach algebra, let $P : A \to \mathbb{C}$ be an orthogonally additive n-homogeneous polynomial, and let $\varphi : A^n \to \mathbb{C}$ be the corresponding symmetric n-linear functional. Suppose $x_1, \ldots, x_k, x_{k+1}, \ldots, x_n \in A$, $1 \le k \le n - 1$, are such that

$$x_i x_j = x_j x_i = 0 \quad (1 \le i \le k, k + 1 \le j \le n). \tag{9.3}$$

Hence

$$\left(\sum_{i=1}^{k} z_i x_i \right) \cdot \left(\sum_{j=k+1}^{n} z_j x_j \right) = \left(\sum_{j=k+1}^{n} z_j x_j \right) \cdot \left(\sum_{i=1}^{k} z_i x_i \right) = 0$$

for any complex numbers z_i. Since P is orthogonally additive, it follows that

$$P\left(\sum_{t=1}^{n} z_t x_t \right) = P\left(\sum_{i=1}^{k} z_i x_i \right) + P\left(\sum_{j=k+1}^{n} z_j x_j \right).$$

As φ is n-linear, we can view the three expressions occurring in this identity as polynomials in $z_1, \ldots, z_n$. Observe that the coefficient at $z_1 \cdots z_n$ on the left-hand side is $n! \varphi(x_1, \ldots, x_n)$, while on the right-hand side is 0. Hence,

$$\varphi(x_1, \ldots, x_n) = 0. \tag{9.4}$$

We have thus shown that an orthogonally additive n-homogeneous polynomial $P(x) = \varphi(x, \ldots, x)$ satisfies the following condition:

$$\text{For every } 1 \le k \le n - 1, (9.3) \text{ implies } (9.4). \tag{9.5}$$

Note that the converse is also true. Thus, P is orthogonally additive if and only if it satisfies (9.5).

What can be said if (9.3) implies (9.4) for a fixed k? In the next theorem, we handle the case where A is a commutative unital zpd Banach algebra and $k = n - 1$ (this is obviously equivalent to the case where $k = 1$, but the $k = n - 1$ case is notationally more convenient). Let us point out that the assumption that φ is symmetric is not needed in this theorem.

Some notation is needed. Let $\varphi : A^n \to \mathbb{C}$ be an n-linear functional, and let I be a subset of $\{1, \ldots, n - 1\}$. If $I \neq \emptyset$, then write $I = \{i_1, \ldots, i_k\}$ with $i_1 < \cdots < i_k$, and define

$$\varphi_I(x_1, \ldots, x_n) = \varphi(\hat{x}_1, \ldots, \hat{x}_{n-1}, x_{i_1} \ldots x_{i_k} x_n),$$

where

$$\hat{x}_j = \begin{cases} x_j, & j \notin I \\ 1, & j \in I. \end{cases}$$

If $I = \emptyset$, then let $\varphi_I = \varphi$.

Theorem 9.11 *Let A be a commutative unital zpd Banach algebra. If $\varphi : A^n \to \mathbb{C}, n \geq 2$, is a continuous n-linear functional such that for all $x_1, \ldots, x_{n-1}, x_n \in A$,*

$$x_1 x_n = \cdots = x_{n-1} x_n = 0 \implies \varphi(x_1, \ldots, x_n) = 0, \tag{9.6}$$

then

$$\sum (-1)^{|I|} \varphi_I(x_1, \ldots, x_n) = 0 \quad (x_1, \ldots, x_n \in A), \tag{9.7}$$

where the sum runs over all subsets I of $\{1, \ldots, n - 1\}$.

Proof If $n = 2$, then (9.7) reads as $\varphi(x_1, x_2) - \varphi(1, x_1 x_2) = 0$; this, of course, is a direct consequence of the assumption that A is zpd. We may thus assume that $n > 2$ and that the theorem is true for continuous $(n - 1)$-linear functionals on A^{n-1}.

Take any $x_2, \ldots, x_{n-1}$ and x_n in A such that $x_2 x_n = \cdots = x_{n-1} x_n = 0$ and define $\psi : A^2 \to \mathbb{C}$ by

$$\psi(x, y) = \varphi(x, x_2, \ldots, x_{n-1}, x_n y).$$

Observe that $xy = 0$ implies

$$x(x_n y) = x_2(x_n y) = \cdots = x_{n-1}(x_n y) = 0,$$

and hence, by our assumption on φ, $\psi(x, y) = 0$. Obviously, ψ is a continuous bilinear functional. Since A is zpd, it follows that $\psi(x, 1) = \psi(1, x)$ for all $x \in A$, that is,

$$\varphi(x, x_2, \ldots, x_{n-1}, x_n) = \varphi(1, x_2, \ldots, x_{n-1}, xx_n) \quad (x \in A).$$

Now fix an arbitrary $x_1 \in A$. Writing x_1 for x, we can restate the above conclusion as that the continuous $(n-1)$-linear functional $\theta : A^{n-1} \to \mathbb{C}$ defined by

$$\theta(x_2, \ldots, x_{n-1}, x_n) = \varphi(x_1, x_2, \ldots, x_{n-1}, x_n) - \varphi(1, x_2, \ldots, x_{n-1}, x_1 x_n)$$

satisfies the condition

$$x_2 x_n = \cdots = x_{n-1} x_n = 0 \implies \theta(x_2, \ldots, x_n) = 0.$$

By our assumption, the assertion of the theorem holds for θ; that is,

$$\sum (-1)^{|I|} \theta_I(x_2, \ldots, x_n) = 0 \quad (x_2, \ldots, x_n \in A),$$

where the sum runs over all subsets I of $\{2, \ldots, n-1\}$. Observe that this identity is nothing but (9.7).   □

It should be pointed out that (9.7) trivially implies (9.6), so the two conditions are actually equivalent. This is because $\varphi_I(x_1, \ldots, x_n)$ involves an expression of the form $x_i x_n$, $i < n$, whenever $I \neq \emptyset$. For example, for $n = 4$, (9.7) reads as

$$\varphi(x_1, x_2, x_3, x_4) = \varphi(1, x_2, x_3, x_1 x_4) + \varphi(x_1, 1, x_3, x_2 x_4) + \varphi(x_1, x_2, 1, x_3 x_4)$$
$$- \varphi(1, 1, x_3, x_1 x_2 x_4) - \varphi(1, x_2, 1, x_1 x_3 x_4) - \varphi(x_1, 1, 1, x_2 x_3 x_4)$$
$$+ \varphi(1, 1, 1, x_1 x_2 x_3 x_4).$$

Modifying the trick used in the proof of Theorem 9.11, we can consider the situations where k is different from 1 and $n - 1$. However, the formulas that can be obtained seem rather messy, even for small n. We will not develop them here. This may be a challenge for the future. Theorem 9.11, which is a new result that has not been published elsewhere, can serve as a sample.

We return briefly to orthogonally additive polynomials. Using Theorem 9.11 for $n = 3$, it is an easy exercise to show that every continuous orthogonally additive 3-homogeneous polynomials on a commutative unital zpd Banach algebra A can be presented in the form (9.2); in fact,

$$P(x) = \varphi(x, x, x) = \varphi(1, 1, x^3) \quad (x \in A).$$

We do not know at present whether this can be extended to an arbitrary n. Unfortunately, Theorem 9.11 is not sufficient for establishing this. For example, if ω is a symmetric bilinear functional on a commutative Banach algebra A, then $\varphi : A^4 \to \mathbb{C}$ defined by

$$\varphi(x_1, x_2, x_3, x_4) = \omega(x_1 x_2, x_3 x_4) + \omega(x_1 x_3, x_2 x_4) + \omega(x_2 x_3, x_1 x_4)$$

is a symmetric 4-linear functional on A that satisfies the condition of Theorem 9.11, but

$$P(x) = \varphi(x, x, , x, x) = 3\omega(x^2, x^2)$$

is not necessarily of the form (9.2). Thus, for larger n, one is forced to tackle the problem indicated in the preceding paragraph.

It is natural to ask what other conditions on a symmetric n-linear functional φ imply that the n-homogeneous polynomial $P(x) = \varphi(x, \ldots, x)$ is of the form (9.2). In the remaining of the section, we will look into one such condition, following the discussion in [16].

Let A be a Banach algebra. An n-linear functional $\varphi \colon A^n \to \mathbb{C}$ is said to be *orthosymmetric* if $\varphi(x_1, \ldots, x_n) = 0$ whenever $x_i x_j = x_j x_i = 0$ for some distinct $i, j \in \{1, \ldots, n\}$ (the name comes from an analogous notion in lattice theory [37]). Note that if A is commutative, then, for any linear functional τ on A,

$$\varphi(x_1, \ldots, x_n) = \tau(x_1 \cdots x_n) \tag{9.8}$$

defines a symmetric orthosymmetric n-linear functional, and the corresponding n-homogeneous polynomial $P(x) = \varphi(x, \ldots, x)$ is of the form (9.2) (with $T = \tau$). Without assuming the commutativity, an obvious example in the $n = 2$ case is given by

$$\varphi(x_1, x_2) = \tau(x_1 \circ x_2).$$

For $n \geq 3$, however, the existence of a nonzero symmetric orthosymmetric n-linear functional depends on the structure of the algebra in question.

The following is [16, Theorem 3.1].

Theorem 9.12 *Let A be a zpd Banach algebra having a bounded approximate identity. If $\varphi \colon A^n \to \mathbb{C}$, $n \geq 2$, is a continuous symmetric orthosymmetric n-linear map, then there exists a $\tau \in A'$ such that*

$$\varphi(x_1, \ldots, x_n) = \tau\big((\ldots((x_1 \circ x_2) \circ x_3) \ldots) \circ x_n\big) \quad (x_1, \ldots, x_n \in A).$$

The proof is rather short; for $n = 2$, the result follows from Theorem 6.1, and then one proceeds by induction on n.

If A is commutative, then Theorem 9.12 can be equivalently stated as that φ is of the desired form (9.8). If A is not commutative and $n \geq 3$, then the expression

$$(\ldots((x_1 \circ x_2) \circ x_3) \ldots) \circ x_n$$

is not symmetric in $x_1, \ldots, x_n$, and so it may happen that the only possibility is that $\varphi = 0$. Specifically, if $n = 3$, then, assuming that A is as in Theorem 9.12, a nonzero continuous symmetric orthosymmetric 3-linear functional $\varphi : A^3 \to \mathbb{C}$ exists if and only if $\overline{[A, A]} \neq A$ [16, Theorem 3.3]. For example, this holds for the algebra $A = M_n(\mathbb{C})$, and in this case φ is of the form

$$\varphi(x, y, z) = \lambda \text{tr}(xyz + zyx)$$

for some $\lambda \in \mathbb{C}$, where $\text{tr}(x)$ stands for the trace of the matrix x [16, Corollary 3.5]. If $n \geq 4$ and A satisfies the weakly Wiener property (e.g., it is unital), then a nonzero continuous symmetric orthosymmetric n-linear functional $\varphi : A^n \to \mathbb{C}$ exists if and only if the closed ideal generated by $[A, A]$ is not equal to A [16, Theorem 3.7].

9.3   Nonassociative Products of Matrices

Let $(A, +, *)$ be a nonassociative algebra. We say that A is *power-associative* if every subalgebra generated by a single element is associative. Such an algebra is, in particular, third power-associative—recall from Sect. 4.1 that this means that

$$(x * x) * x = x * (x * x) \quad (x \in A).$$

Next, we say that A is *Lie-admissible* if it is a Lie algebra under the product

$$\{x, y\} = x * y - y * x.$$

Of course, associative algebras are Lie-admissible. More generally, if A is endowed with another associative product $(x, y) \mapsto xy$ and

$$x * y - y * x = xy - yx, \tag{9.9}$$

then $(A, +, *)$ is Lie-admissible.

In [34], Benkart and Osborn proved that if A is the linear space $M_n(F)$ of $n \times n$ matrices over F, with $n \geq 2$ and F a field of characteristic not 2 or 3, and $*$ is a power-associative product on $M_n(F)$ satisfying (9.9), where xy is the standard product of matrices, then there exist a scalar α, a linear functional μ on $M_n(F)$, and a symmetric bilinear functional σ on $M_n(F)$ such that

$$x * y = \alpha xy + (\alpha - 1)yx + \mu(x)y + \mu(y)x + \sigma(x, y)1 \quad (x, y \in M_n(F))$$

(here, 1 stands for the identity matrix). Beidar and Chebotar [32] noticed that by applying the theory of functional identities, this can be extended to much more general algebras (see also [56, Section 8.1]).

Combining functional identities with the fact that $M_n(F)$ is a zLpd algebra, we will now extend the theorem by Benkart and Osborn in a different direction. Specifically, instead of assuming (9.9) we will only assume that $xy = yx$ if and only if $x * y = y * x$ (i.e., $[x, y] = 0$ if and only if $\{x, y\} = 0$). The results we will present are taken from [53].

To avoid misunderstanding, we point out that we will consider the linear space $M_n(F)$ equipped with two products: the standard product xy and the "unknown" nonassociative product $x * y$. That is, $(x, y) \mapsto x * y$ is just a bilinear map from $M_n(F) \times M_n(F)$ to $M_n(F)$. By $[x, y]$ (respectively, $x \circ y$) we denote the usual Lie (respectively, Jordan) product of $x, y \in M_n(F)$.

Throughout this section, we assume that F is a field. As usual, by $\mathrm{sl}_n(F)$ we denote the Lie algebra of trace zero matrices in $M_n(F)$.

We start with a simple application of the fact that $M_n(F)$ is a zLpd algebra (Corollary 3.11).

Lemma 9.13 *Let $*$ be a nonassociative product on the linear space $M_n(F)$. The following two conditions are equivalent:*

(i) *For all $x, y \in M_n(F)$, $xy=yx$ implies $x * y = y * x$.*
(ii) *There exists a linear map $T : \mathrm{sl}_n(F) \to M_n(F)$ such that*

$$x * y - y * x = T([x, y]) \quad (x, y \in M_n(F)).$$

Proof Assume (i). Applying Corollary 3.11, together with Proposition 1.3 (iv), to the bilinear map

$$(x, y) \mapsto x * y - y * x,$$

we see that (ii) holds. The converse is trivial. $\qquad\square$

Remark 9.14 Note that any commutative product $*$ satisfies condition (i). Assuming that the characteristic of F is not 2, we have

$$x * y = \frac{1}{2}(x * y - y * x) + \frac{1}{2}(x * y + y * x), \tag{9.10}$$

and so condition (ii) can be written

$$x * y = S([x, y]) + x \odot y,$$

where $S = \frac{T}{2}$ and $x \odot y = \frac{1}{2}(x * y + y * x)$ is a commutative product. Perhaps this gives a clearer picture of the lemma.

Under the assumption that the converse implication in (i) also holds, Lemma 9.13 obtains the following form.

Lemma 9.15 *Let $*$ be a nonassociative product on the linear space $M_n(F)$. The following two conditions are equivalent:*

(i) *For all $x, y \in M_n(F)$, $xy = yx$ if and only if $x * y = y * x$.*
(ii) *There exists a bijective linear map $T : M_n(F) \to M_n(F)$ such that*

$$x * y - y * x = T([x, y]) \quad (x, y \in M_n(F)).$$

Proof It is clear that (ii) implies (i), so we only have to prove the converse. Assuming (i), we see from Lemma 9.13 that there exists a linear map $T : \mathrm{sl}_n(F) \to M_n(F)$ such that $x * y - y * x = T([x, y])$ for all $x, y \in M_n(F)$. It is well-known that every element in $\mathrm{sl}_n(F)$ can be expressed as a commutator $[x, y]$ [23]. Since (i) states that $T([x, y]) = x * y - y * x = 0$ implies $[x, y] = 0$, we see that T has trivial kernel. Hence, we can extend T to a bijective linear map from $M_n(F)$ to itself. This proves (ii). ☐

The proofs of the next two theorems use functional identities, which we briefly discussed in Sect. 7.2 on commutativity preserving linear maps. More precisely, we will need two basic results that concern bilinear maps $B : A \times A \to A$ that satisfy functional identities that are similar, or even the same, as those to which we arrived when considering commutativity preservers. Although these results hold for considerably more general algebras, we now state them, for simplicity of exposition, only for the case where $A = M_n(F)$, $n \geq 2$.

(a) If the characteristic of F is not 2 and B satisfies

$$[B(x, x), x] = 0 \quad (x \in M_n(F)),$$

then there exist a scalar λ, a linear functional μ on $M_n(F)$, and a bilinear functional ν on $M_n(F)$ such that

$$B(x, x) = \lambda x^2 + \mu(x)x + \nu(x, x)1 \quad (x \in M_n(F)).$$

(This follows from [56, Theorem 5.32 and Remark 5.33].)
(b) If $n \geq 3$, B is skew-symmetric and satisfies

$$[B(x, y), z] + [B(z, x), y] + [B(y, z), x] = 0 \quad (x, y, z \in M_n(F)),$$

then there exist a scalar λ, a linear functional μ on $M_n(F)$, and a skew-symmetric bilinear functional v on $M_n(F)$ such that

$$B(x, y) = \lambda[x, y] + \mu(x)y - \mu(y)x + v(x, y)1 \quad (x, y \in M_n(F)).$$

(This follows by combining three important results on functional identities. The first one is [56, Corollary 2.23], which tells us that $M_n(F)$ is 3-free (here we used that $n \neq 2$), the second one is [56, Corollary 4.14], which tells us that B is a quasi-polynomial, and the third one is [56, Lemma 4.4], which easily yields the final conclusion.)

In the proof of the first theorem, we will apply (a).

Theorem 9.16 *Let $*$ be a nonassociative product on the linear space $M_n(F)$. If the characteristic of F is not 2, then the following two conditions are equivalent:*

(i) *For all $x, y \in M_n(F)$, $xy = yx$ if and only if $x * y = y * x$, and $*$ is third power-associative.*
(ii) *There exist a bijective linear map $S : M_n(F) \to M_n(F)$, a scalar α, a linear functional τ on $M_n(F)$, and a symmetric bilinear functional σ on $M_n(F)$ such that*

$$x * y = S([x, y]) + \alpha x \circ y + \tau(x)y + \tau(y)x + \sigma(x, y)1 \quad (x, y \in M_n(F)).$$

Proof It is straightforward to check that (ii) implies (i). Assume (i). Lemma 9.15 shows that there is a bijective linear map $T : M_n(F) \to M_n(F)$ such that

$$x * y - y * x = T([x, y]) \quad (x, y \in M_n(F)).$$

Writing $x * x$ for y and using the assumption that $*$ is third power-associative, it follows that $[x, x * x] = 0$ for every $x \in M_n(F)$. We can now use (a) to conclude that there exist a scalar λ, a linear functional μ on $M_n(F)$, and a bilinear functional v on $M_n(F)$ such that

$$x * x = \lambda x^2 + \mu(x)x + v(x, x)1 \quad (x \in M_n(F)).$$

Replacing x by $x + y$, we obtain

$$x * y + y * x = \lambda(x \circ y) + \mu(x)y + \mu(y)x + \big(v(x, y) + v(y, x)\big)1 \quad (x, y \in M_n(F)).$$

Writing $x * y$ as in (9.10) and setting

$$S = \frac{T}{2}, \quad \alpha = \frac{\lambda}{2}, \quad \tau = \frac{\mu}{2}, \quad \sigma(x, y) = \frac{1}{2}\big(v(x, y) + v(y, x)\big),$$

we see that (ii) holds.   □

The proof of the second theorem is based on (b).

Theorem 9.17 *Let* $*$ *be a nonassociative product on the linear space* $M_n(F)$. *If* $n \geq 3$, *then the following two conditions are equivalent:*

(i) *For all* $x, y \in M_n(F)$, $xy = yx$ *if and only if* $x * y = y * x$, *and* $(M_n(F), +, *)$ *is a Lie-admissible algebra.*

(ii) *There exist a scalar* λ *and a linear functional* γ *on* $\mathrm{sl}_n(F)$ *such that*

$$x * y - y * x = \lambda[x, y] + \gamma([x, y])1 \quad (x, y \in M_n(F)).$$

Moreover, $\lambda z \neq -\gamma(z)1$ *for every nonzero* $z \in \mathrm{sl}_n(F)$.

Proof Showing that (ii) implies (i) is again straightforward. Thus, assume that (i) holds. As in the preceding proof, we apply Lemma 9.15 to obtain a bijective linear map $T :$ $M_n(F) \to M_n(F)$ such that

$$\{x, y\} = T([x, y]) \quad (x, y \in M_n(F)), \tag{9.11}$$

where

$$\{x, y\} = x * y - y * x.$$

Since the algebra $(M_n(F), +, *)$ is Lie-admissible, the Jacobi identity holds for $\{\cdot, \cdot\}$, that is,

$$\{\{x, y\}, z\} + \{\{z, x\}, y\} + \{\{y, z\}, x\} = 0 \quad (x, y, z \in M_n(F)).$$

We can rewrite this as

$$T\big([\{x, y\}, z] + [\{z, x\}, y] + [\{y, z\}, x]\big) = 0 \quad (x, y, z \in M_n(F)).$$

As T is bijective, this gives

$$[\{x, y\}, z] + [\{z, x\}, y] + [\{y, z\}, x] = 0 \quad (x, y, z \in M_n(F)).$$

We are now in a position to use (b). Thus,

$$\{x, y\} = \lambda[x, y] + \mu(x)y - \mu(y)x + \nu(x, y)1 \quad (x, y \in M_n(F)) \tag{9.12}$$

for some scalar λ, a linear functional μ on $M_n(F)$, and a skew-symmetric bilinear functional ν on $M_n(F)$.

Let e_{ij} denote the usual matrix units in $M_n(F)$. If μ was not 0, there would exist $i, j \in \{1, \ldots, n\}$ such that $\mu(e_{ij}) \neq 0$. As $n \geq 3$, we can take $k \in \{1, \ldots, n\}$ different from i and j. Therefore, $[e_{ij}, e_{kk}] = 0$, and hence, by our assumption, $\{e_{ij}, e_{kk}\} = 0$. However, this is impossible in light of (9.12). Therefore, $\mu = 0$.

From (9.11) and (9.12), we see that there is a linear functional γ on $\mathrm{sl}_n(F)$ such that $\nu(x, y) = \gamma([x, y])$. By our assumption, $[x, y] \neq 0$ implies $\{x, y\} \neq 0$. Again using the fact that every element in $\mathrm{sl}_n(F)$ is a commutator [23], we see that this implies that $\lambda z + \gamma(z)1 \neq 0$ for every nonzero $z \in \mathrm{sl}_n(F)$. $\qquad\square$

Example 9.18 The assumption in Theorem 9.17 that $n \neq 2$ is necessary. Indeed, denoting by x^t the transpose of the matrix x, one can verify that the product

$$x * y = x^t y^t$$

satisfies condition (i) but is not of the form described in (ii).

Remark 9.19 Assuming that the characteristic of F is not 2, we can, similarly as in Remark 9.14, restate condition (ii) of Theorem 9.17 as

$$x * y = \beta[x, y] + \delta([x, y])1 + x \odot y,$$

where $x \odot y$ is a commutative product, $\beta = \frac{\lambda}{2}$ is a scalar, and $\delta = \frac{\gamma}{2}$ is a linear functional. Moreover, if the characteristic of F is 0 or does not divide n, then the condition that $\lambda z \neq -\gamma(z)1$ for every nonzero trace zero matrix z is obviously automatically fulfilled as long as $\lambda \neq 0$.

Combining Theorems 9.16 and 9.17, we finally obtain a generalization of the result by Benkart and Osborn.

Corollary 9.20 *Let $*$ be a nonassociative product on the linear space $M_n(F)$. If $n \geq 3$ and the characteristic of F is not 2, then the following two conditions are equivalent:*

(i) *For all $x, y \in M_n(F)$, $xy = yx$ if and only if $x * y = y * x$, and $(M_n(F), +, *)$ is a third power-associative Lie-admissible algebra.*

(ii) *There exist scalars λ_1 and λ_2, a linear functional τ on $M_n(F)$, a linear functional δ on $\mathrm{sl}_n(F)$, and a symmetric bilinear functional σ on $M_n(F)$ such that*

$$x * y = \alpha_1 xy + \alpha_2 yx + \tau(x)y + \tau(y)x + \big(\sigma(x, y) + \delta([x, y])\big)1 \quad (x, y \in M_n(F)).$$

Moreover, $(\alpha_1 - \alpha_2)z \neq -2\delta(z)1$ for every nonzero $z \in \mathrm{sl}_n(F)$.

Proof It is enough to prove that (i) implies (ii). Assuming (i), Theorem 9.16 tells us that

$$\frac{1}{2}(x * y + y * x) = \alpha x \circ y + \tau(x)y + \tau(y)x + \sigma(x, y)1 \quad (x, y \in M_n(F)),$$

and Theorem 9.17 tells us that

$$\frac{1}{2}(x * y - y * x) = \frac{\lambda}{2}[x, y] + \frac{\gamma}{2}([x, y])1 \quad (x, y \in M_n(F)).$$

Adding these two identities, we arrive at the desired conclusion, with $\alpha_1 = \alpha + \frac{\lambda}{2}$, $\alpha_2 = \alpha - \frac{\lambda}{2}$, and $\delta = \frac{\gamma}{2}$. Since

$$x * y - y * x = (\alpha_1 - \alpha_2)[x, y] + 2\delta([x, y]) \quad (x, y \in M_n(F)),$$

we see, as at the end of the proof of Theorem 9.17, that $(\alpha_1 - \alpha_2)z \neq -2\delta(z)1$ for every nonzero $z \in \mathrm{sl}_n(F)$.   □

Remark 9.21 If the characteristic of F is 0 or does not divide n, then $(\alpha_1 - \alpha_2)z \neq -2\delta(z)1$ for every nonzero $z \in \mathrm{sl}_n(F)$ is equivalent to $\alpha_1 \neq \alpha_2$.

It is plausible that the above results can be extended to more general algebras. However, we will not go into this here. Like elsewhere in Part III, our goal was primarily to present the methods based on the zpd concept and thereby indicate how the theory discussed in this book can be used in "real life."

References

1. K. Aceves, M. Dugas, Local multiplication maps on $F[x]$. J. Algebra Appl. **14**, 1550029 (2015)
2. B.A.E. Ahlem, A.M. Peralta, Linear maps on C^*-algebras which are derivations or triple derivations at a point. Linear Algebra Appl. **538**, 1–21 (2018)
3. J. Alaminos, M. Brešar, A.R. Villena, The strong degree of von Neumann algebras and the structure of Lie and Jordan derivations. Math. Proc. Cambridge Philos. Soc. **137**, 441–463 (2004)
4. J. Alaminos, M. Brešar, J. Extremera, A.R. Villena, Characterizing homomorphisms and derivations on C^*-algebras. Proc. Roy. Soc. Edinburgh Sect. A **137**, 1–7 (2007)
5. J. Alaminos, M. Brešar, M. Černe, J. Extremera, A.R. Villena, Zero product preserving maps on $C^1[0, 1]$. J. Math. Anal. Appl. **347**, 472–481 (2008)
6. J. Alaminos, M. Brešar, J. Extremera, A.R. Villena, Maps preserving zero products. Studia Math. **193**, 131–159 (2009)
7. J. Alaminos, M. Brešar, J. Extremera, A.R. Villena, Characterizing Jordan maps on C^*-algebras through zero products. Proc. Edinb. Math. Soc. **53**, 543–555 (2010)
8. J. Alaminos, M. Brešar, J. Extremera, A.R. Villena, On bilinear maps determined by rank one idempotents. Linear Algebra Appl. **432**, 738–743 (2010)
9. J. Alaminos, J. Extremera, A.R. Villena, Zero product preserving maps on Banach algebras of Lipschitz functions. J. Math. Anal. Appl. **369**, 94–100 (2010)
10. J. Alaminos, J. Extremera, A.R. Villena, Approximately zero-product-preserving maps. Israel J. Math. **178**, 1–28 (2010)
11. J. Alaminos, J. Extremera, A.R. Villena, Hyperreflexivity of the derivation space of some group algebras. Math. Z. **266**, 571–582 (2010)
12. J. Alaminos, J. Extremera, A.R. Villena, Hyperreflexivity of the derivation space of some group algebras, II. Bull. London Math. Soc. **44**, 323–335 (2012)
13. J. Alaminos, M. Brešar, J. Extremera, Š. Špenko, A.R. Villena, Determining elements in C^*-algebras through spectral properties. J. Math. Anal. Appl. **405**, 214–219 (2013)
14. J. Alaminos, M. Brešar, J. Extremera, Š. Špenko, A.R. Villena, Derivations preserving quasinilpotent elements. Bull. Lond. Math. Soc. **46**, 379–384 (2014)
15. J. Alaminos, J. Extremera, A.R. Villena, Orthogonality preserving linear maps on group algebras. Math. Proc. Cambridge Philos. Soc. **158**, 493–504 (2015)
16. J. Alaminos, M. Brešar, J. Extremera, Š. Špenko, A.R. Villena, Orthogonally additive polynomials and orthosymmetric maps in Banach algebras with properties $\mathbb{A}$ and $\mathbb{B}$. Proc. Edinb. Math. Soc. **59**, 559–568 (2016)
17. J. Alaminos, M. Brešar, J. Extremera, Š. Špenko, A.R. Villena, Commutators and square-zero elements in Banach algebras. Q. J. Math. **67**, 1–13 (2016)

© The Author(s), under exclusive license to Springer Nature Switzerland AG 2021
M. Brešar, *Zero Product Determined Algebras*, Frontiers in Mathematics,
https://doi.org/10.1007/978-3-030-80242-4

18. J. Alaminos, M. Brešar, J. Extremera, A.R. Villena, Zero Lie product determined Banach algebras. Studia Math. **239**, 189–199 (2017)

19. J. Alaminos, M. Brešar, J. Extremera, A.R. Villena, Zero Lie product determined Banach algebras, II. J. Math. Anal. Appl. **474**, 1498–1511 (2019)

20. J. Alaminos, M. Brešar, J. Extremera, A.R. Villena, Zero Jordan product determined Banach algebras. J. Austral. Math. Soc. to appear. doi:10.1017/S1446788719000478

21. J. Alaminos, J. Extremera, M.L.C. Godoy, A.R. Villena, Orthogonally additive polynomials on convolution algebras associated with a compact group. J. Math. Anal. Appl. **472**, 285–302 (2019)

22. J. Alaminos, M.L.C. Godoy, A.R. Villena, Orthogonally additive polynomials on the algebras of approximable operators. Linear Multilinear Algebra **67**, 1922–1936 (2019)

23. A.A. Albert, B. Muckenhoupt, On matrices of trace zero. Michigan Math. J. **4**, 1–3 (1957)

24. G. An, J. Li, J. He, Zero Jordan product determined algebras. Linear Algebra Appl. **475**, 90–93 (2015)

25. J. Araujo, K. Jarosz, Biseparating maps between operator algebras. J. Math. Anal. Appl. **282**, 48–55 (2003)

26. W. Arendt, Spectral properties of Lamperti operators. Indiana Univ. Math. J. **32**, 199–215 (1983)

27. B. Aupetit, *A Primer on Spectral Theory*. Universitext (Springer, New York, 1991)

28. S. Banach, *Théorie des Opérations Linéaires*, Warsaw (1932)

29. W.E. Baxter, W.S. Martindale 3rd, Jordan homomorphisms of semiprime rings. J. Algebra **56**, 457–471 (1979)

30. E. Beckenstein, L. Narici, The separating map: a survey. Rend. Circ. Mat. Palermo **52**, 637–648 (1998)

31. K.I. Beidar, M. Brešar, Extended Jacobson density theorem for rings with derivations and automorphisms. Israel J. Math. **122**, 317–346 (2001)

32. K.I. Beidar, M.A. Chebotar, On Lie-admissible algebras whose commutator Lie algebras are Lie subalgebras of prime associative algebras. J. Algebra **233**, 675–703 (2000)

33. H. Behncke, Nilpotent elements in group algebras. Bull. Acad. Polon. Sci. Sér. Sci. Math. Astronom. Phys. **19**, 197–198 (1971)

34. G. Benkart, J.M. Osborn, Power-associative products on matrices. Hadronic J. Math. **5**, 1859–1892 (1982)

35. D. Benkovič, M. Grašič, Generalized derivations on unital algebras determined by action on zero products. Linear Algebra Appl. **445**, 347–368 (2014)

36. Y. Benyamini, S. Lassalle, J.G. Llavona, Homogeneous orthogonally additive polynomials on Banach lattices. Bull. London Math. Soc. **38**, 459–469 (2006)

37. K. Boulabiar, G. Buskes, Vector lattice powers: f-algebras and functional calculus. Comm. Algebra **34**, 1435–1442 (2006)

38. A. Bourhim, J. Mashreghi, A. Stepanyan, Maps between Banach algebras preserving the spectrum. Arch. Math. **107**, 609–621 (2016)

39. G. Braatvedt, R. Brits, Uniqueness and spectral variation in Banach algebras. Quaest. Math. **36**, 155–165 (2013)

40. G. Braatvedt, R. Brits, H. Raubenheimer, Spectral characterizations of scalars in a Banach algebra. Bull. Lond. Math. Soc. **41**, 1095–1104 (2009)

41. M. Brešar, Jordan mappings of semiprime rings. J. Algebra **127**, 218–228 (1989)

42. M. Brešar, Characterizations of derivations on some normed algebras with involution. J. Algebra **152**, 454–462 (1992)

43. M. Brešar, Jordan derivations revisited. Math. Proc. Cambridge Philos. Soc. **139**, 411–425 (2005)

44. M. Brešar, Characterizing homomorphisms, derivations and multipliers in rings with idempotents. Proc. Royal Soc. Edinb. Sect. A **137**, 9–21 (2007)
45. M. Brešar, Jordan homomorphisms revisited. Math. Proc. Cambridge Philos. Soc. **144**, 317–328 (2008)
46. M. Brešar, Multiplication algebra and maps determined by zero products. Linear Multilinear Algebra **60**, 763–768 (2012)
47. M. Brešar, *Introduction to Noncommutative Algebra*. Universitext (Springer, Berlin, 2014)
48. M. Brešar, Finite dimensional zero product determined algebras are generated by idempotents. Expo. Math. **34**, 130–143 (2016)
49. M. Brešar, Functional identities and zero Lie product determined Banach algebras. Q. J. Math. **71**, 649–665 (2020)
50. M. Brešar, Commutators and images of noncommutative polynomials. Adv. Math. **374**, 107346, 21 (2020)
51. M. Brešar, M. Mathieu, Derivations mapping into the radical, III. J. Funct. Anal. **133**, 21–29 (1995)
52. M. Brešar, P. Šemrl, On locally linearly dependent operators and derivations. Trans. Amer. Math. Soc. **351**, 1257–1275 (1999)
53. M. Brešar, P. Šemrl, On bilinear maps on matrices with applications to commutativity preservers. J. Algebra **301**, 803–837 (2006)
54. M. Brešar, P. Šemrl, Zero product determined maps on matrix rings over division rings. Contemporary Math. **750**, 195–213 (2020)
55. M. Brešar, Š. Špenko, Determining elements in Banach algebras through spectral properties. J. Math. Anal. Appl. **393**, 144–150 (2012)
56. M. Brešar, M.A. Chebotar, W.S. Martindale 3rd, *Functional Identities*. Frontiers in Mathematics (Birkhäuser, Basel, 2007)
57. M. Brešar, E.Kissin, V. Shulman, Lie ideals: from pure algebra to C^*-algebras. J. Reine Angew. Math. **623**, 73–121 (2008)
58. M. Brešar, M. Grašič, J. Sanchez, Zero product determined matrix algebras. Linear Algebra Appl. **430**, 1486–1498 (2009)
59. M. Brešar, B. Magajna, Š. Špenko, Identifying derivations through the spectra of their values. Integral Equ. Oper. Theory **73**, 395–411 (2012)
60. M. Brešar, X. Guo, G. Liu, R. Lü, K. Zhao, Zero product determined Lie algebras. European J. Math. **5**, 424–453 (2019)
61. D. Brice, H. Huang, On zero product determined algebras. Linear Multilinear Algebra **63**, 326–342 (2015)
62. R. Brits, F. Schulz, C. Touré, A spectral characterization of isomorphisms on C^*-algebras. Arch. Math. **113** , 391–398 (2019)
63. G. Brown, W. Moran, Point derivations on $M(G)$. Bull. London Math. Soc. **8**, 57–64 (1976)
64. M. Burgos, F. J. Fernández-Polo, A. M. Peralta, Local triple derivations on C^*-algebras and JB^*-triples. Bull. Lond. Math. Soc. **46**, 709–724 (2014)
65. D. Carando, S. Lassalle, I. Zalduendo, Orthogonally additive polynomials over $C(K)$ are measures—a short proof. Integral Equ. Oper., Theory **56**, 597–602 (2006)
66. L. Catalano, On maps characterized by action on equal products. J. Algebra **511**, 148–154 (2018)
67. M.A. Chebotar, W.-F. Ke, P.-H. Lee, N.-C. Wong, Mappings preserving zero products. Studia Math. **155**, 77–94 (2003)
68. M.A. Chebotar, W.-F. Ke, P.-H. Lee, Maps characterized by action on zero products. Pacific J. Math. **216**, 217–228 (2004)
69. M.A. Chebotar, W.-F. Ke, P.-H. Lee, Maps preserving zero Jordan products on Hermitian operators. Illinois J. Math. **49**, 445–452 (2005)

70. M.A. Chebotar, W.-F. Ke, P.-H. Lee, R. Zhang, On maps preserving zero Jordan products. Monatsh. Math. **149**, 91–101 (2006)
71. M.A. Chebotar, P.-H. Lee, E. Puczylowski, On commutators and nilpotent elements in simple rings. Bull. Lond. Math. Soc. **42**, 191–194 (2010)
72. Y. Chen, J. Li, Mappings on some reflexive algebras characterized by action on zero products or Jordan zero products. Studia Math. **206**, 121–134 (2011)
73. L. Chen, F. Lu, T. Wang, Local and 2-local Lie derivations of operator algebras on Banach spaces. Integral Equ. Oper. Theory **77**, 109–121 (2013)
74. P.M. Cohn, The range of derivations on a skew field and the equation $ax - xb = c$. J. Indian Math. Soc. **37**, 61–69 (1973)
75. R.L. Crist, Local derivations on operator algebras. J. Funct. Anal. **135**, 76–92 (1996)
76. H.G. Dales, *Banach Algebras and Automatic Continuity*. London Mathematical Society Monographs, New Series, vol. 24 (Oxford Science Publications, The Clarendon Press, Oxford University Press, New York, 2000)
77. C. de Seguins Pazzis, A note on sums of three square-zero matrices. Linear Multilinear Algebra **65**, 787–805 (2017)
78. M. Dugas, B. Wagner, Finitary incidence algebras and idealizations. Linear Multilinear Algebra **64**, 1936–1951 (2016)
79. J. Duncan, A.W. Tullo, Finite dimensionality, nilpotents and quasinilpotents in Banach algebras. Proc. Edinburgh Math. Soc. **19**, 45–49 (1974/1975)
80. E. Erdogan, Ö. Gök, Convolution factorability of bilinear maps and integral representations. Indag. Math. **29**, 1334–1349 (2018)
81. E. Erdogan, Ö. Gök, E.A. Sánchez Pérez, Product factorability of integral bilinear operators on Banach function spaces. Positivity **23**, 671–696 (2019)
82. E. Erdogan, E.A. Sánchez Pérez, Integral representation of product factorable bilinear operators and summability of bilinear maps on $C(K)$-spaces. J. Math. Anal. Appl. **483**, 123629, 25 (2020)
83. P.A. Fillmore, Sums of operators with square zero. Acta Sci. Math. **28**, 285–288 (1967)
84. H. Ghahramani, Zero product determined triangular algebras. Linear Multilinear Algebra **61**, 741–757 (2013)
85. H. Ghahramani, On derivations and Jordan derivations through zero products. Oper. Matrices **8**, 759–771 (2014)
86. S. Goldstein, Stationarity of operator algebras. J. Funct. Anal. **118**, 275–308 (1993)
87. M. Grašič, Zero product determined classical Lie algebras. Linear Multilinear Algebra **58**, 1007–1022 (2010)
88. M. Grašič, Zero product determined Jordan algebras, I. Linear Multilinear Algebra **59**, 741–757 (2011)
89. M. Grašič, Zero product determined Jordan algebras, II. Algebra Colloq. **22**, 109–118 (2015)
90. N. Grønbæk, Weak and cyclic amenability for noncommutative Banach algebras. Proc. Edinburgh Math. Soc. **35**, 315–328 (1992)
91. U. Haagerup, N.J. Laustsen, Weak amenability of C^*-algebras and a theorem of Goldstein, in *Banach Algebras '97* (Blaubeuren) (de Gruyter, Berlin, 1998), pp. 223–243
92. D. Hadwin, J. Li, Local derivations and local automorphisms. J. Math. Anal. Appl. **290**, 702–714 (2004)
93. B. Harris, Commutators in division rings. Proc. Amer. Math. Soc. **9**, 628–630 (1958)
94. I.N. Herstein, Jordan homomorphism. Trans. Amer. Math. Soc. **81**, 331–341 (1956)
95. I.N. Herstein, Jordan derivations of prime rings. Proc. Amer. Math. Soc. **8**, 1104–1110 (1957)
96. I.N. Herstein, Lie and Jordan structures in simple associative rings. Bull. Amer. Math. Soc. **67**, 517–531 (1961)

97. C. Heunen, C. Horsman, Matrix multiplication is determined by orthogonality and trace. Linear Algebra Appl. **439**, 4130–4134 (2013)

98. J. Hou, L. Zhao, Zero-product preserving additive maps on symmetric operator spaces and self-adjoint operator spaces. Linear Algebra Appl. **399**, 235-244 (2005)

99. J. Hou, L. Zhao, Jordan zero-product preserving additive maps on operator algebras. J. Math. Anal. Appl. **314**, 689–700 (2006)

100. W. Hu, Z. Xiao, A characterization of algebras generated by idempotents. J. Pure Appl. Algebra **225**, 106693 (2021)

101. N. Jacobson, C. Rickart, Jordan homomorphisms of rings. Trans. Amer. Math. Soc. **69**, 479–502 (1950)

102. P. Ji, W. Qi, Characterizations of Lie derivations of triangular algebras. Linear Algebra Appl. **435**, 1137–1146 (2011)

103. W. Jing, S. Lu, P. Li, Characterization of derivations on some operator algebras. Bull. Austral. Math. Soc. **66**, 227–232 (2002)

104. B.E. Johnson, Symmetric amenability and the nonexistence of Lie and Jordan derivations. Math. Proc. Cambridge Philos. Soc. **120**, 455–473 (1996)

105. B.E. Johnson, Local derivations on C^*-algebras are derivations. Trans. Amer. Math. Soc. **353**, 313–325 (2001)

106. B.E. Johnson, A.M. Sinclair, Continuity of derivations and a problem of Kaplansky. Amer. J. Math. **90**, 1067–1073 (1968)

107. R.V. Kadison, Isometries of operator algebras. Ann. Math. **54**, 325–338 (1951)

108. R.V. Kadison, Local derivations. J. Algebra **130**, 494–509 (1990)

109. R.V. Kadison, J.R. Ringrose, *Fundamentals of the Theory of Operator Algebras. Vol. I.* Pure and Applied Mathematics, vol. 100 (Academic, New York, 1983)

110. E. Kaniuth, *A Course in Commutative Banach Algebras.* Graduate Texts in Mathematics, vol. 246 (Springer, New York, 2009)

111. E. Kaniuth, A.T.-M. Lau, *Fourier and Fourier-Stieltjes Algebras on Locally Compact Groups.* Mathematical Surveys and Monographs, vol. 231 (American Mathematical Society, Providence, 2018)

112. I. Kaplansky, Dual rings. Ann. Math. **49**, 689–701 (1948)

113. N. Kataoka, Finite sums of nilpotent elements in properly infinite C^*-algebras. Hokkaido Math. J. **31**, 275–281 (2002)

114. A. Katavolos, C. Stamatopoulos, Commutators of quasinilpotents and invariant subspaces. Studia Math. **128**, 159–169 (1998)

115. M.T. Kosan, T.-K. Lee, Y. Zhou, Bilinear forms on matrix algebras vanishing on zero products of xy and yx. Linear Algebra Appl. **453**, 110–124 (2014)

116. T.Y. Lam, *A First Course in Noncommutative Rings.* Graduate Texts in Mathematics, 2nd edn. (Springer, Berlin, 2001)

117. J. Lamperti, On the isometries of certain functions spaces. Pacific J. Math. **8**, 459–466 (1958)

118. D.R. Larson, A.R. Sourour, Local derivations and local automorphisms of $B(X)$, in *Proceedings of Symposia in Pure Mathematics*, vol. 51, Part 2 (American Mathematical Society, Providence, 1990), pp. 187–194

119. A.T.-M. Lau, N.-C. Wong, Orthogonality and disjointness preserving linear maps between Fourier and Fourier-Stieltjes algebras of locally compact groups. J. Funct. Anal. **265**, 562–593 (2013)

120. T.-K. Lee, Generalized skew derivations characterized by acting on zero products. Pacific J. Math. **216**, 293–301 (2004)

121. T.-K. Lee, J.-H. Lin, Jordan derivations of prime rings with characteristic two. Linear Algebra Appl. **462**, 1–15 (2014)

122. C. Le Page, Sur quelques conditions entraînant la commutativité dans les algèbres de Banach. C. R. Acad. Sc. Paris Sér. A-B **265**, A235–A237 (1967)

123. H. Li, Y. Liu, Maps completely preserving commutativity and maps completely preserving Jordan zero-product. Linear Algebra Appl. **462**, 233–249 (2014)

124. Y.-F. Lin, Completely bounded disjointness preserving operators between Fourier algebras. J. Math. Anal. Appl. **382**, 469–473 (2011)

125. C.-K. Liu, P.-K. Liau, Generalized derivations preserving quasinilpotent elements in Banach algebras. Linear Multilinear Algebra **66**, 1888–1908 (2018)

126. F. Lu, W. Jing, Characterizations of Lie derivations of $B(X)$. Linear Algebra Appl. **432**, 89–99 (2010)

127. X. Ma, G. Ding, L. Wang, Square-zero determined matrix algebras. Linear Multilinear Algebra **59**, 1311–1317 (2011)

128. M. Mackey, Local derivations on Jordan triples. Bull. Lond. Math. Soc. **45**, 811–824 (2013)

129. B. Magajna, On the relative reflexivity of finitely generated modules of operators. Trans. Amer. Math. Soc. **327**, 221–249 (1991)

130. L. Marcoux, Linear span of projections in certain simple C^*-algebras. Indiana Univ. Math. J. **51**, 753–771 (2002)

131. L. Marcoux, Sums of small number of commutators. J. Oper. Th. **56**, 111–142 (2006)

132. M. Mathieu, Where to find the image of a derivation. Functional analysis and operator theory, polish academy of sciences. Banach Cent. Publ. **30**, 237–249 (1994)

133. M. Mathieu, G. Murphy, Derivations mapping into the radical. Arch. Math. **57**, 469–474 (1991)

134. L. Molnár, *Selected Preserver Problems on Algebraic Structures of Linear Operators and on Function Spaces*. Lecture Notes in Mathematics (Springer, Berlin, 2007)

135. P. Moravec, Unramified Brauer groups of finite and infinite groups. Amer. J. Math. **134**, 1679–1704 (2012)

136. G. Murphy, Aspects of the theory of derivations. functional analysis and operator theory, polish academy of sciences. Banach Cent. Publ. **30**, 267–275 (1994)

137. W.K. Nicholson, Lifting idempotents and exchange rings. Trans. Amer. Math. Soc. **229**, 269–278 (1977)

138. A. Nowicki, On local derivations in the Kadison sense. Colloq. Math. **89**, 193–198 (2001)

139. C. Palazuelos, A.M. Peralta, I. Villanueva, Orthogonally additive polynomials on C^*-algebras. Q. J. Math. **59**, 363–374 (2008)

140. C. Pearcy, D. Topping, Sums of small numbers of idempotents. Michigan Math. J. **14**, 453–465 (1967)

141. C. Pearcy, D. Topping, Commutators and certain II_1-factors. J. Funct. Anal. **3**, 69–78 (1969)

142. C. Pearcy, D. Topping, On commutators in ideals of compact operators. Michigan J. Math. **18**, 247–252 (1971)

143. D. Pérez-García, I. Villanueva, Orthogonally additive polynomials on spaces of continuous functions. J. Math. Anal. Appl. **306**, 97–105 (2005)

144. V. Pták, Derivations, commutators and the radical. Manuscripta Math. **23**, 355–362 (1977/1978)

145. C.J. Read, Discontinuous derivations on the algebra of bounded operators on a Banach space. J. London Math. Soc. **40**, 305–326 (1989)

146. L. Robert, On the Lie ideals of C^*-algebras. J. Oper. Theory **75**, 387–408 (2016)

147. S. Roch, P.A. Santos, B. Silbermann, *Non-commutative Gelfand Theories. A Tool-kit for Operator Theorists and Numerical Analysts*. Universitext (Springer, Berlin, 2011)

148. E. Samei, Approximately local derivations. J. London Math. Soc. **71**, 759–778 (2005)

149. E. Samei, Reflexivity and hyperreflexivity of bounded n-cocycles from group algebras. Proc. Amer. Math. Soc. **139**, 163–176 (2011)

150. E. Samei, J. Soltan Farsani, Hyperreflexivity of bounded N-cocycle spaces of Banach algebras. Monatsh. Math. **175**, 429–455 (2014)

151. E. Samei, J. Soltan Farsani, Hyperreflexivity constants of the bounded n-cocycle spaces of group algebras and C^*-algebras. J. Aust. Math. Soc. **109**, 112–130 (2020)

152. F. Schulz, R. Brits, Uniqueness under spectral variation in the socle of a Banach algebra. J. Math. Anal. Appl. **444**, 1626–1639 (2016)

153. P. Šemrl, Local automorphisms and derivations on $B(H)$. Proc. Amer. Math. Soc. **125**, 2677–2680 (1997)

154. V.S. Shulman, Operators preserving ideals in C^*-algebras. Studia Math. **109**, 67–72 (1994)

155. A.M. Sinclair, Continuous derivations on Banach algebras. Proc. Amer. Math. Soc. **20**, 166–170 (1969)

156. I.M. Singer, J. Wermer, Derivations on commutative normed algebras. Math. Ann. **129**, 260–264 (1955)

157. J. Soltan Farsani, Hyperreflexivity of the bounded n-cocycle spaces of Banach algebras with matrix representations. Studia Math. **250**, 35–55 (2020)

158. N. Stopar, Preserving zeros of XY and XY^*. Comm. Algebra **40**, 2053–2065 (2012)

159. K. Sundaresan, Geometry of spaces of homogeneous polynomials on Banach lattices, in *Applied Geometry and Discrete Mathematics*. DIMACS: Series in Discrete Mathematics and Theoretical Computer Science, vol. 4 (American Mathematical Society, Providence, 1991), pp. 571–586

160. M.P. Thomas, The image of a derivation is contained in the radical. Ann. Math. **128**, 435–460 (1988)

161. C. Touré, F. Schulz, R. Brits, Truncation and spectral variation in Banach algebras. J. Math. Anal. Appl. **445**, 23–31 (2017)

162. Yu.V. Turovskii, V.S. Shulman, Conditions for the massiveness of the range of a derivation of a Banach algebra and of associated differential operators. Mat. Zametki **42**, 305–314 (1987)

163. A.R. Villena, Orthogonally additive polynomials on Banach function algebras. J. Math. Anal. Appl. **448**, 447–472 (2017)

164. J.H. Wang, P.Y. Wu, Sums of square-zero operators. Studia Math. **99**, 115–127 (1991)

165. D. Wang, X. Yu, Z. Chen, A class of zero product determined Lie algebras. J. Algebra **331**, 145–151 (2011)

166. D. Wang, X. Yu, Z. Chen, Maps determined by action on square-zero elements. Comm. Algebra **40**, 4255–4262 (2012)

167. W. Watkins, Linear maps that preserve commuting pairs of matrices. Linear Algebra Appl. **14**, 29–35 (1976)

168. M. Wolff, Disjointness preserving operators on C^*-algebras. Arch. Math. (Basel) **62**, 248–253 (1994)

169. W.J. Wong, Maps on simple algebras preserving zero products. I. The associative case. Pacific J. Math. **89**, 229–247 (1980)

Index